Introduction to Stochastic Analysis and Malliavin Calculus

Introduction to Stochastic Analysis and Malliavin Calculus

Editor

Jai Rathod

Introduction to Stochastic Analysis and Malliavin Calculus

Edited by **Jai Rathod**

Printed in 2017

ISBN: 978-1-68117-190-6

Library of Congress Control Number: 2015949134

© 2016 by
SCITUS Academics LLC,
616, Corporate Way, Suite 2, 4766,
Valley Cottage, NY 10989

www.scitusacademics.com

This book contains information obtained from highly regarded resources. Copyright for individual articles remains with the authors as indicated. All chapters are distributed under the terms of the Creative Commons Attribution License, which permits unrestricted use, distribution, and reproduction in any medium, provided the original author and source are credited.

Notice

Reasonable efforts have been made to publish reliable data and views articulated in the chapters are those of the individual contributors, and not necessarily those of the editors or publishers. Editors or publishers are not responsible for the accuracy of the information in the published chapters or consequences of their use. The publisher believes no responsibility for any damage or grievance to the persons or property arising out of the use of any materials, instructions, methods or thoughts in the book. The editors and the publisher have attempted to trace the copyright holders of all material reproduced in this publication and apologize to copyright holders if permission has not been obtained. If any copyright holder has not been acknowledged, please write to us so we may rectify.

Preface

Stochastic calculus is a branch of mathematics that operates on stochastic processes. It allows a consistent theory of integration to be defined for integrals of stochastic processes with respect to stochastic processes. It is used to model systems that behave randomly. The best-known stochastic process to which stochastic calculus is applied is the Wiener process, the Wiener process has been widely applied in financial mathematics and economics to model the evolution in time of stock prices and bond interest rates.

The Malliavin calculus extends the calculus of variations from functions to stochastic processes. The Malliavin calculus is also called the stochastic calculus of variations. In particular, it allows the computation of derivatives of random variables. Malliavin's ideas led to a proof that Hörmander's condition implies the existence and smoothness of a density for the solution of a stochastic differential equation; Hörmander's original proof was based on the theory of partial differential equations. The calculus has been applied to stochastic partial differential equations as well. The calculus allows integration by parts with random variables; this operation is used in mathematical finance to compute the sensitivities of financial derivatives. The calculus has applications in, for example, stochastic filtering. This book emphasizes on differential stochastic equations and Malliavin calculus.

Table of Contents

Simplifying Stochastic Mathematical Models of Biochemical Systems 1

The Existence and Uniqueness of Random Solution to Itô Stochastic Integral Equation 23

Solving Nonlinear Stochastic Diffusion Models With Nonlinear Losses Using The Homotopy Analysis Method 39

Stochastic Process Optimization Technique 65

Two Implicit Runge-Kutta Methods for Stochastic Differential Equation 87

Parameter Dependence in Stochastic Modeling—Multivariate Distributions 105

Set-Valued Stochastic Integrals with Respect to Finite Variation Processes 133

A New Formula For Partitions In A Set Of Entities Into Empty And Nonempty Subsets, And Its Application To Stochastic And Agent-Based Computational Models 149

Upwind Finite-Volume Solution Of Stochastic Burgers' Equation 167

Solution of Nonlinear Stochastic Langevin's Equation Using WHEP, Pickard and HPM Methods 191

Index 221

Simplifying Stochastic Mathematical Models of Biochemical Systems

Silvana Ilie and Samaneh Gholami
Department of Mathematics, Ryerson University,
Toronto, Canada

ABSTRACT

Stochastic modeling of biochemical reactions taking place at the cellular level has become the subject of intense research in recent years. Molecular interactions in a single cell exhibit random fluctuations. These fluctuations may be significant when small populations of some reacting species are present and then a stochastic description of the cellular dynamics is required. Often, the biochemically reacting systems encountered in applications consist of many species interacting through many reaction channels. Also, the dynamics of such systems is typically non-linear and presents multiple time-scales. Consequently, the stochastic mathematical models of biochemical systems can be quite complex and their analysis challenging. In this paper, we present a method to reduce a stochastic continuous model of well-stirred biochemical systems, the Chemical Langevin Equation, while preserving the overall behavior of the system. Several tests of our method on models of practical interest gave excellent results.

INTRODUCTION

Mathematical modeling of biochemical reactions within a cell is crucial for understanding cellular dynamics [1]. Biological processes are often represented as systems of chemical reactions. In general, such processes involve many reacting species subject to many reaction channels. In systems with some low molecular populations, the random fluctuations associated with molecular interactions may be significant, as was observed experimentally [2, 3]. Then, a deterministic model is not appropriate and stochastic models are needed for an accurate description of the reacting system behavior [4]. In many applications, the biochemical reactions evolve on widely different time scales. The presence of the multiple time-scale dynamics leads to mathematical stiffness. Stiffness poses great challenges in the simulation of the system dynamics. Moreover, many biochemical systems are non-linear which makes them more difficult to analyze. Thus, the stochastic mathematical models of biochemically reacting systems can be quite complex and their simulation computationally very demanding [5].

Often the goal is to understand some specific biological process of a complex biochemical network. Then, one needs to identify a reduced system of reactions which gives an accurate description of that process. Model reduction strategies may focus on reducing the number of reactions, the number of parameters or that of molecular species [6]. The existing model reduction schemes for deterministic models of chemical reaction systems may be grouped into sensitivity analysis methods [7], lumping methods [8,9] and time-scale analysis methods [10,11]. Lumping techniques lead to loss of information about particular species or reactions and thus the physical interpretation of the elementary reactions is lost. They may be appropriate when only limited information is available about specific reactions. Time-scale analysis applies to systems with rapid and slow reactions channels and is based on the assumption that the fast components are in a quasi-steady state. The slow dynamics is restricted to the algebraic constraints defining the equilibrium manifold of the fast components. The identification of the fast and slow reactions is required as well as estimations of the orders of magnitude of separation between them. Finally, sensitivity analysis may be employed to eliminate the reactions which are not important, if the parametric sensitivity with respect to their reaction rate constants are small. Then the physical insight offered by the elementary reactions, as opposed to group of reactions, is maintained. Much less work has been dedicated to designing model reduction strategies for the stochastic models of biochemical kinetics (see [12-14] and references therein).

Our contribution in this paper is to provide a novel method for reducing a stochastic continuous model of biochemical systems, the Chemical Langevin Equation.

The Chemical Langevin Equation (CLE) model [15] is an approximation of the discrete stochastic model of well-stirred biochemical kinetics, namely the Chemical Master Equation [16-19]. The CLE model is valid in the regime of large molecular numbers. The CLE is a system of Itô stochastic differential equations (SDE) with multiplicative noise of dimension equal to the number of reacting species. Our method is based on sensitivity analysis, which is a key tool for modeling and analyzing biochemical systems. Sensitivity analysis is widely used in quantifying the characteristics of the system [20], such as robustness with respect to perturbations in its parameters. These parameters include the reaction rate constants or the initial supplies of species. The study of the dependence of the system dynamics on its parameters is a critical problem in modeling biochemical systems, and in particular cellular dynamics, as usually some parameters for the kinetics of interaction are unknown or cannot be measured with accuracy. Also, in cellular environments, the supply of reactants may fluctuate. Sensitivity analysis studies the variation in molecular populations with respect to small changes in parameters. Thus, it enables the identification of the kinetic parameters with a negligible impact on the species of interest. Then, the reactions corresponding to these parameters can be removed from the system. By this procedure, the dynamics of interest in not altered while the system may be significantly reduced. Moreover, the reduced system may provide important physical insight and it is easier to analyze and simulate numerically.

The outline of the paper is as follows. Section 2 gives an introduction to the stochastic continuous modeling of well-stirred biochemical system. In Section 3, a technique for estimating sensitivities of the Chemical Langevin Equation is explored. In Section 4, we provide a model reduction strategy for the Chemical Langevin Equation, and we test this method on several models of interest in Section 5.

CHEMICAL LANGEVIN EQUATION

We present below a brief introduction to the stochastic modeling of chemical kinetics for well-stirred systems, in the regime of large molecular populations. The system is at thermal equilibrium in a constant volume. The chemically reacting system contains N biochemical species S_1 ,S_N subject to M reactions R_1, ,R_M. Under the assumptions above, the system dynamics can be described by the vector of states X(t), with component $X_i(t)$ being the number of S_i molecules available at time t, for each i=1, ,N. The dynamical state X(t) is a Markov process.

Let us define v_{ij} to be the change in the number of molecules of species S_i caused by one reaction R_j and denote by $v_j = (v_{1j},...,v_{Nj})'$ the state-change vector characterizing the reaction R_j. The matrix $S = \{v_{ij}\}_{1 \leq i \leq N, 1 \leq j \leq M}$ is known as the stoichiometric matrix. Remark that each reaction R_j produces a change in the system given by $x \to x + v_j$. In representing the reaction R_j an essential role is played by propensity function $a_j(x)$. The propensity function is defined by $a_j(x)dt$ at the given state x, is the probability that one reaction R_j will happen in the infinitesimal timeinterval [t,t+dt]. For example, for a reaction $S_1 \xrightarrow{c_1} S_2$ the propensity function is $a_1(x) = c_1 x_1$, while for a reaction $S_1 + S_2 \xrightarrow{c_2} S_3$ the propensity function is $a_2(x) = c_2 x_1 x_2$. Finally, the associated propensity of the reaction $S_1 + S_2 \xrightarrow{c_3} s_4$ is $a_3(x) = c_3 x_1 (x_1 - 1)/2$.

In addition, the following assumptions are made: there exists a timestep $\tau > 0$ such that 1) the step τ is small enough so that there is no significant change in any propensity during the interval [t, t+ τ],

$$a_j(X(s)) \approx a_j(X(t)) \text{ for } s \in [t,t+\tau] \tag{1}$$

Simplifying Stochastic Mathematical Models of Biochemical Systems

for each j=1, ,M and 2) the step ⊠ is sufficiently large so that the expected number of times each reaction R_j fires during $[t, t+\tau]$ is large. That is, each propensity should satisfy

$$a_j(X(t))\tau \gg 1 \text{ for } j = 1, \cdots, M. \tag{2}$$

Under the conditions above, the state vector obeys

$$dX(t) = \sum_{j=1}^{M} v_j a_j(X(t))dt + \sum_{j=1}^{M} v_j \sqrt{a_j(X(t))} dW_j(t) \tag{3}$$

Where W_j are independent Wiener processes for j=1, ,M The Equation (3) is known as the Chemical Langevin Equation. This model is obtained when the requirements of 1) and 2) are satisfied. These conditions hold when each species has large molecular populations. The CLE is a system of Itô stochastic differential equations with multiplicative non-commutative noise, having one equation for each reacting species. The solution of the CLE model (3) should satisfy the initial condition

$$X(0) = x_0 \tag{4}$$

imposed at the initial time t=0.

Applying the expectation (denoted by $\mathbb{E}$) to the Chemical Langevin Equation (3) leads to

$$\begin{aligned} &d(\mathbb{E}(X)) \\ &= \sum_{j=1}^{M} v_j \mathbb{E}(a_j(X(t)))dt + \mathbb{E}\left(\sum_{j=1}^{M} v_j \sqrt{a_j(X(t))} dW_j(t)\right) \\ &= \sum_{j=1}^{M} v_j \mathbb{E}(a_j(X(t)))dt. \end{aligned}$$

Now, dividing by dt, yields

$$\frac{d}{dt}(\mathbb{E}(X)) = \sum_{j=1}^{M} v_j \mathbb{E}(a_j(X(t))).$$

Notice that the CLE model reduces to the classical reaction rate equation model of the chemical kinetics, when all species have very large molecular numbers, in thermodynamic limit,

$$\frac{dX(t)}{dt} = \sum_{j=1}^{M} v_j a_j(X(t)). \tag{5}$$

The reaction rate equation is a system of ordinary differential equations of dimension equal to the number of biochemical species in the system. Generally, the reaction rate Equation (5) is written in terms of concentrations rather than in molecular population numbers.

PARAMETRIC SENSITIVITY OF THE CHEMICAL LANGEVIN EQUATION

Biochemical reaction models may depend on many parameters such as the kinetic rates, the initial amounts for each species or an uncertain environment. Some small changes in the parameters may considerably affect the system behavior. Hence, it is important to determine the influences of such changes. Sensitivity analysis studies the dependence of the system dynamics on the reaction rate parameters or the initial conditions. It is an essential analysis tool in kinetic modeling and it may be used to decide which parts of the model are actively contributing to the system behavior. Therefore, it plays a key role in assessing the accuracy of a model, in analyzing the model and in model reduction. In general, if X is differentiable with respect to a parameter p, the first order sensitivity is defined as $\partial X / \partial p$. A large sensitivity suggests that the system may change dramatically when that parameter is perturbed. This shows that an accurate measurement of that parameter is necessary. By contrast, a small sensitivity indicates that the system is robust with respect to small variations in that parameter and thus rough estimations of the parameter value will be sufficient.

Simplifying Stochastic Mathematical Models of Biochemical Systems

Below we discuss a numerical technique for computing the pathwise (strong) sensitivities with respect to the kinetic parameters for the Chemical Langevin Equation model. Pathwise derivative estimation of stochastic discrete models was studied for biochemical systems [21,22] as well for other areas of applications [23,24]. Since the CLE is a system of SDEs, then sensitivity analysis tools available for SDEs may apply for the Langevin model. The pathwise sensitivity technique presented here applies to diffusion processes [23].

Let $X(t,p,\omega) = (x_1, x_2, \cdots, x_N)'(t,p,\omega)$ be the solution of the CLE model (3) satisfying the initial condition (4). Here ω is a realization of the sample trajectories space Ω and p is a kinetic parameter. The pathwise sensitivity is interpreted as the pathwise derivative of $X(t,p,\omega)$ with respect to $p, (\partial X/\partial p)(t,p,\omega)$, when the realization ω is fixed. The pathwise sensitivities may be obtained by differentiating, on each trajectory, the Chemical Langevin Equation (3) with respect to the parameter of interest, p. If the derivative $(\partial X/\partial p)(t,p,\omega)$ exists with probability 1, we derive

$$d\left(\frac{\partial X}{\partial p}\right)$$

$$= \sum_{k=1}^{M} v_k \left[\frac{\partial a_k(X)}{\partial X}\frac{\partial X}{\partial p} + \frac{\partial a_k(X)}{\partial p}\right](t)dt$$

$$+ \sum_{k=1}^{M} v_k \left[\frac{1}{2\sqrt{a_k(X)}}\left(\frac{\partial a_k(X)}{\partial X}\frac{\partial X}{\partial p} + \frac{\partial a_k(X)}{\partial p}\right)\right](t)dW_k$$

(6)

Notice that v_k does not depend on p for all $1 \leq k \leq M$, while the propensities are polynomials of degree at most one in the kinetic parameters thus $\partial a_k(X)/\partial p$ does not depend on p explicitly. This approach to computing the path wise sensitivities may be applied when each molecular species is bounded away from zero, which is true for biochemical systems in the Langevin regime.

In order to calculate the local sensitivities, the coupled system consisting of the CLE (3) and the auxiliary equations for the sensitivities (6) is solved to find the state vector X(t) and the sensitivities $\partial X(t)/\partial p$. Remark that the combined system (3) and (6) has double size compared to the Chemical Langevin Equation, but has the same number of independent Wiener processes. Hence, it is generally almost twice as expensive to solve numerically as the CLE. The pathwise sensitivity analysis uses the exact derivative with respect to a parameter instead of numerical differentiation, as does a finite-difference scheme. It can be estimated along with the solution of the CLE on each individual Brownian path, independently of the other paths.

Note that we are only interested in the sensitivity with respect to the reaction rate constants. The initial amounts of molecules are independent of the kinetic parameters, and thus the initial conditions for the sensitivities obey

$$\frac{\partial X}{\partial p}(0) = 0. \tag{7}$$

In the following, this analysis will be used to reduce the complexity of the biochemical reaction system. Note that the parametric sensitivity presented here is accurate for choices of the kinetic parameters in some neighborhood of the values for which the analysis was applied, since our analysis focuses on estimating the local sensitivities.

A MODEL REDUCTION STRATEGY

We provide below a novel strategy for reducing the complexity of stochastic continuous models of wellstirred biochemical kinetics. The aim is to reduce the original model to a smaller one which preserves the dynamics, the stability properties and the physical relevance of the full system. This approach extends that of a parametric sensitivity for continuous deterministic systems, and in particular for the reaction rate equations [7].

Simplifying Stochastic Mathematical Models of Biochemical Systems

Our method utilizes the pathwise (strong) sensitivity analysis for diffusive processes in the case of the Chemical Langevin Equation. Only the sensitivity with respect to the kinetic parameters is considered. The strategy will help identify the parameters having a strong influence of the system dynamics. The parameters which have an insignificant impact on the overall behavior the biochemical system, or on the species of interest, are then eliminated. Hence, their reactions are deleted, which leads also to the elimination of the unimportant species. This is particulary important, as the Chemical Langevin Equation model has as many equations as reacting species. Since the deleted species are constant in the reduced model, then the associated reduced CLE model becomes lower dimensional. These species are set to their initial value. Moreover, the Wiener processes associated with the deleted reactions are also removed, simplifying the numerical simulation and improving its efficiency.

For the Chemical Langevin model, the non-dimensional pathwise sensitivities depend on the particular Brownian path considered. Denote by $\varepsilon \ll 1$ the tolerance and by $\omega \in \Omega$ a realization in the sample trajectory space. To estimate the pathwise sensitivity of $X(t,p,w)$ with respect to the parameter $p, \partial X/\partial p(t,p,\omega)$, for a fixed sample trajectory $\omega \in \Omega$, we integrate numerically the system (3) and (6), with the initial conditions (4) and (7) on that trajectory. Then, we impose the following strong criterion: the CLE system is robust (not sensitive) with respect to the parameter p if the non-dimensional (scaled) sensitivities satisfy

$$\left| \frac{p}{X_i(t,\omega)} \frac{\partial X_i}{\partial p}(t,\omega) \right| < \varepsilon,$$

for all $\omega \in \Omega$ and all $0 \leq t \leq T$, \hfill (8)

for all $1 \leq i \leq N$ or all species of interest. For the purpose of applying the sensitivity analysis to simplify stochastic continuous models of biochemical kinetics, this strong criterion may not be necessary.

We propose the following requirement for the elimination of the unimportant reactions, which is a weaker criterion than (8), but quite useful for simplifying biochemical kinetics. A reaction with reaction rate p may be eliminated, if for a given tolerance $\varepsilon \ll 1$, the sensitivities with respect to p obey

$$\frac{p}{|\mathbb{E}(X_i(t))|}\left(\mathbb{E}\left(\left|\frac{\partial X_i}{\partial p}(t)\right|\right) + 2 \cdot \sqrt{\mathbb{V}ar\left(\frac{\partial X_i}{\partial p}(t)\right)}\right) < \varepsilon \tag{9}$$

for any $0 \leq t \leq T$ and for all species of interest X_i. Here $\mathbb{V}ar$ denotes the variance.

NUMERICAL RESULTS

In this section, our strategy for simplifying the Chemical Langevin Equation model is tested on several examples of practical interest. In our tests, the Chemical Langevin Equation and the associated system of local sensitivities with respect to each kinetic parameter are integrated numerically. The underlying numerical method is the EulerMaruyama scheme. The reaction rates for which the associated sensitivities in the species of interest are very small, that is satisfy the requirement (9) for some $\varepsilon \ll 1$, are identified. Then, the reactions corresponding to these kinetic parameters are eliminated. Finally, the relative errors in the mean and standard deviation of the reduced compared to the full system are estimated.

Modified Cycle Test Model

First, we study a modified cycle biochemical model [25], undergoing the following reaction channels

$$S_1 \xrightarrow{c_1} S_2, \quad S_2 \xrightarrow{c_2} S_3, \quad S_3 \xrightarrow{c_3} S_1,$$
$$S_1 + S_4 \xrightarrow{c_4} S_5, \quad S_5 \xrightarrow{c_5} S_1 + S_4.$$

Simplifying Stochastic Mathematical Models of Biochemical Systems

The state change vectors of this biochemical system are

$$v_1' = (-1,-1,1)', \quad v_2' = (1,1,-1)',$$
$$v_3' = (-1,1,-1)', \quad v_4' = (1,-1,1)'$$

and $v_5 = (-1,1,1)'$. Furthermore, the reaction propensities are given by

$$a_1(X) = c_1 X_1, \; a_4(X) = c_4 X_1 X_4, \; a_2(X) = c_2 X_2, \; a_5(X) = c_5 X_5, \; a_3(X) = c_3 X_3,$$

with the values of the kinetic parameters being

$$c_1 = 1.5 \times 10^3, \; c_2 = 5 \times 10^3, \; c_3 = 10^3, \; c_4 = 1.66 \times 10^{-3}$$

and $c_5 = 8 \times 10^{-2}$. The model is integrated with initial conditions $X(0) = (1200, 800, 1500, 500, 200)'$ on the time-interval [0, 10]. In our tests, the simulations are done over 10,000 trajectories.

We apply the pathwise sensitivity analysis to this model. The plots of the time-evolution of the sensitivities with respect to the species of interests, S_1, S_2 and S_3, are presented in Figure 1. Notice that the non-dimensional sensitivities with respect to the parameters c_4 and c_5 of the species S_1, S_2 and S_3 are very small, below $2 \times 10^{-2} \ll 1$, thus the system is robust with respect to variations in these parameters. In addition, this shows that the reactions R_4 and R_5, together with the unimportant species S_4 and S_5, can be eliminated from the system, with a negligible effect on the dynamics of the important species, S_1, S_2 and S_3. Figure 2 shows the graphs of the relative error, between the biochemical system without the reactions R_4 and R_5 and the full reaction system, for the mean and standard deviations in the species S_1, S_2 and S_3. The relative errors in the mean are below 2×10^{-2}, while the relative errors in the standard deviation are at most 4×10^{-2}. Consequently, the model reduction technique provided, based on pathwise sensitivity analysis, gives a very accurate representation of the statistical properties of interest for the biochemical system under consideration.

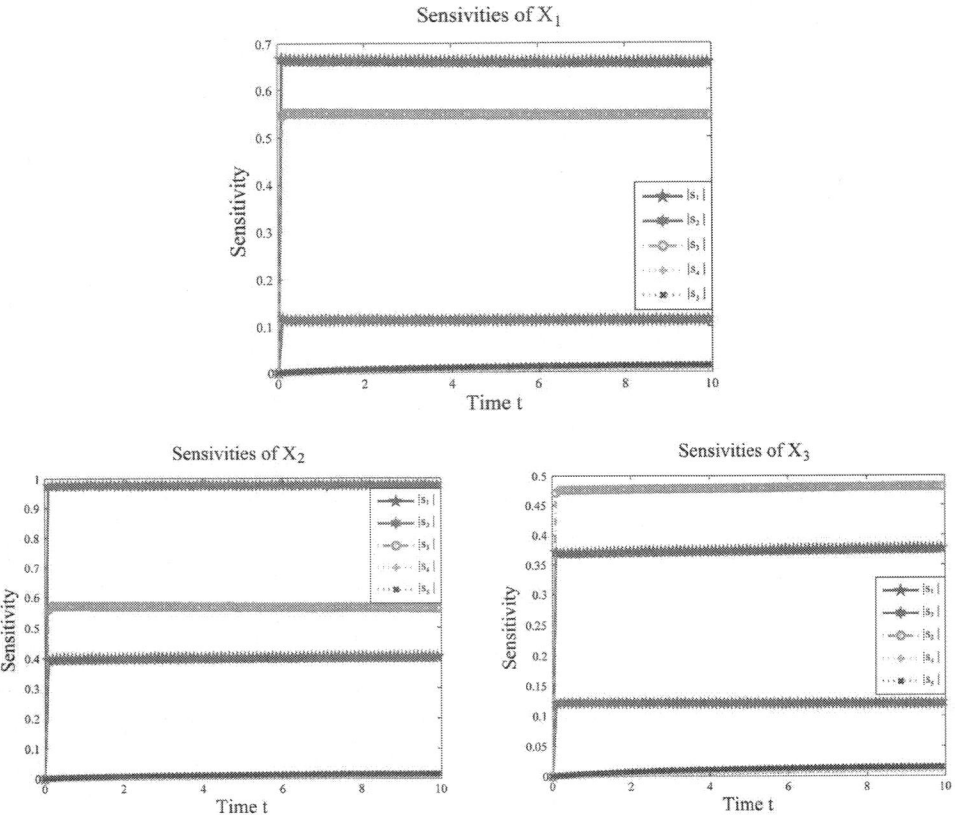

Figure 1: Sensitivity comparison for the cycle test model: the evolution in time of the sensitivities of X_1, X_2 and X_3 with respect to the kinetic parameters, on the interval [0, 10]. The simulation uses 10,000 paths.

Simplifying Stochastic Mathematical Models of Biochemical Systems

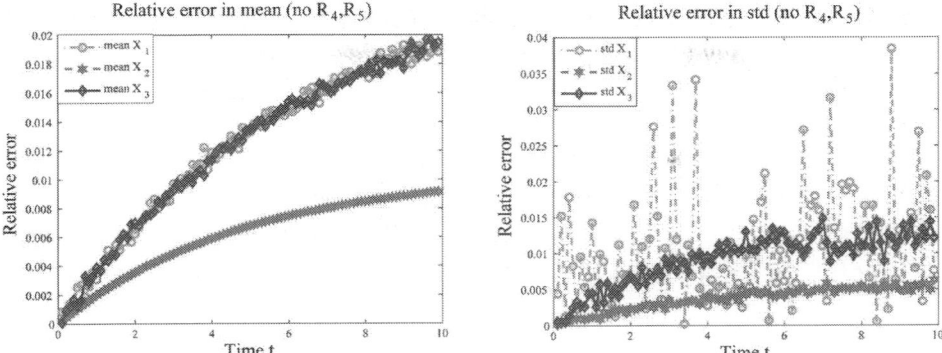

Figure 2: Cycle test model: the evolution in time of the relative errors in means (left) and std (right) of X_1, X_2 and X_3 on the interval [0, 10], when the reactions R_4 and R_5 are eliminated, compared to the full system. The numerical simulations of the solution of the Chemical Langevin Equation used 10,000 trajectories.

Infectious Disease Model

Now, let us consider an infectious disease model [26] involving two components, the species S_1 which caries the infectious disease and the species S_2, which may become infected. The species are subject to the following reactions

$$S_1 \xrightarrow{c_1} \varnothing, \; S_2 \xrightarrow{c_2} \varnothing, \; \varnothing \xrightarrow{c_3} S_1,$$
$$\varnothing \xrightarrow{c_4} S_2, \; S_1 + S_2 \xrightarrow{c_5} S_1 + S_1.$$

The reactions have state-change vectors given by $v_1 = -v_3 = (-1,0)'$, $v_2 = -v_4 = (0,-1)'$ and $v_5 = (1,-1)'$. Each reactions is characterized by a propensity, expressed as

$$a_1(X) = c_1 X_1, \; a_4(X) = c_4, \; a_2(X) = c_2 X_2,$$
$$a_5(X) = c_5 X_1 X_2, \; a_3(X) = c_3.$$

For this model, the kinetic parameters are $c_1 = 2.0, c_2 = 0.1, c_3 = 25, c_4 = 75$ and $c_5 = 0.05$. Integration is performed on the time interval [0, 10], with the initial conditions $X(0) = (20, 40)'$. For our simulations, we approximate the exact solution of the Chemical Langevin Equation on 10,000 trajectories.

The pathwise sensitivity method for the Chemical Langevin Equation model is employed for this model. Figure 3 depicts the evolution of the non-dimensional sensitivities of the species S_1 and S_2 with respect to the kinetic parameters. Remark that the sensitivities with respect to c_2 are quite small, less than 7×10^{-2} for species S_1 and S_2. This indicates that the reaction R_2 is unimportant, and thus can be deleted, with negligible influence on the overall behavior of the system. Figure 4 presents the evolution of the relative errors in the mean and standard deviation, between the reduced system (with the reaction R_2 removed) and the full system, for the species S_1 and S_2, respectively. The relative errors are below 5×10^{-2}, with the error in the mean being slightly larger than that in the standard deviation. We conclude that the model reduction techniques described above gives very good results on the infectious disease model with the set of parameters considered above.

A Multiscale Biochemical Model

Our final example is a biochemical model [27] undergoing the following reaction channels

$$\emptyset \xrightarrow{c_1} S_1, \ S_1 \xrightarrow{c_2} S_2, \ S_2 \xrightarrow{c_3} S_1,$$
$$S_2 \xrightarrow{c_4} S_3, \ S_3 \xrightarrow{c_5} \emptyset.$$

The propensities associated with the reactions above are while the state-change vectors are

$$v_1 = (1, 0, 0)', v_2 = -v_3 = (-1, 1, 0)', v_4 = (0, -1, 1)'$$

and $v_5 = (0, 0, -1)'$. The kinetic parameters take the values $c_1 = 100, c_2 = c_3 = 150, c_4 = 0.02$ and $c_5 = 0.5$. The system, integrated

Simplifying Stochastic Mathematical Models of Biochemical Systems

on the interval $[0,10]$ with initial conditions $X(0) = (100,100,100)'$, presents multiple scales in time.

We start by employing the sensitivity analysis method described above. The evolution of the non-dimensional sensitivities of the species of interest, S_1 and S_2, with respect to all parameters is depicted in Figure 5. Figure 5 shows that these species are robust with respect to variations in the parameters c_4 and c_5. Indeed, their sensitivities are below $5 \times 10^{-2} \ll 1$.

$$a_1(X) = c_1, \; a_4(X) = c_4 X_2, \; a_2(X) = c_2 X_1,$$
$$a_5(X) = c_5 X_3, \; a_3(X) = c_3 X_2,$$

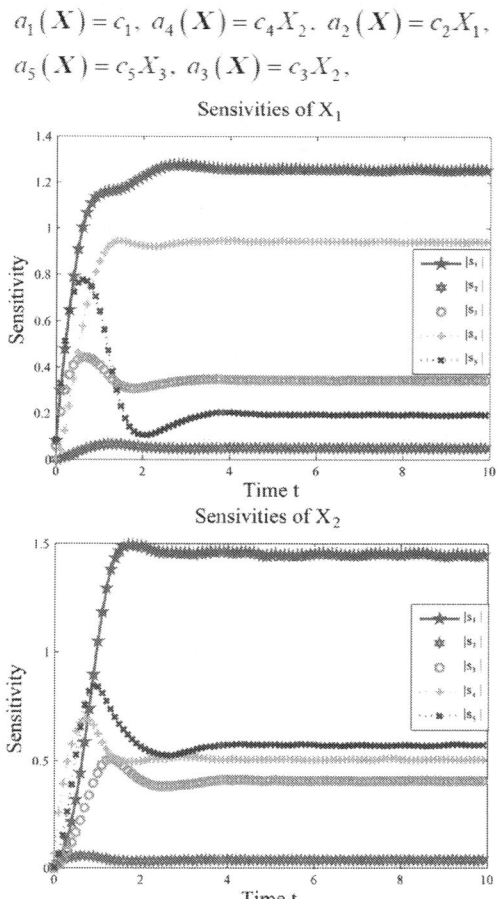

Figure 3: Sensitivity comparison for the infectious disease model: the evolution in time of the non-dimensional sensitivities of X_1 and X_2 with respect to the kinetic parameters, on the interval [0,10]. The simulations are over 10,000 trajectories.

Thus, we can eliminate the reactions R_4 and R_5 and the unimportant species X_3. To estimate the changes in the important species dynamics, produced by the deletion of the unimportant reactions R_4 and R_5, we compute the relative error between the biochemical system with the reactions R_4 and R_5 removed and the full reaction system. The time-evolution of the relative errors in the mean and the standard deviation for the species of interest X_1 and X_2 is plotted in Figure 6.

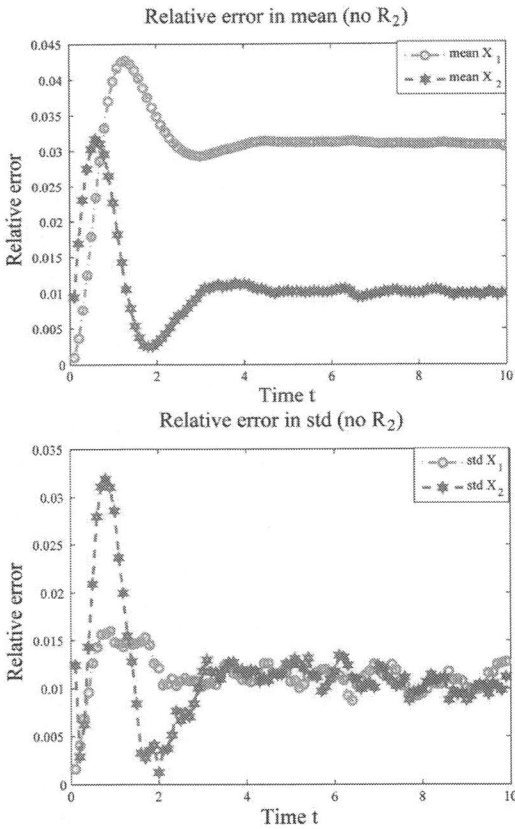

Figure 4: Infectious disease model: the evolution in time of the relative errors in mean and std of X_1 and X_2 in the interval [0,10], when the reaction R_2 is eliminated, compared to the full system. The numerical solution of the Chemical Langevin Equation is computed on 10,000 trajectories.

Simplifying Stochastic Mathematical Models of Biochemical Systems

The relative errors in the mean for the species X_1 and X_2 are below 6×10^{-2}, while the relative errors in the standard deviation are below 3×10^{-2}. The CLE system was reduced from a model with 3 species subject to 5 reactions to a system of 2 species undergoing 3 reactions. We note that our model reduction strategy works very well for this biochemical model with multiple scales in time.

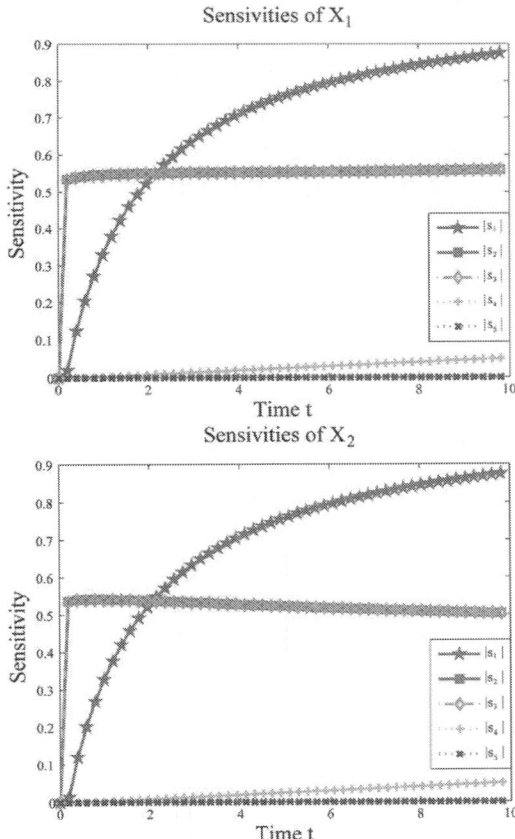

Figure 5: Sensitivity comparison for the multiscale biochemical model: the evolution in time of the non-dimensional sensitivities of X_1 and X_2 with respect to the kinetic parameters, on the interval [0,10]. The simulations are done on 10,000 trajectories.

CONCLUSIONS

In this paper, we developed a method to reduce biochemical systems modeled with the Chemical Langevin Equation. The Chemical Langevin Equation is a system of stochastic differential equations with multiplicative and non-commutative noise. Generally, the biochemical systems arising in applications are quite complex, involving many species and many reactions, and thus are difficult to simulate numerically and to analyze. The model reduction strategy provided utilizes the pathwise sensitivity analysis of stochastic differential equations to identify the parameters with an insignificant influence on the biochemical system dynamics. These parameters are eliminated together with the unimportant species, leading to a smaller model which maintains the characteristics of the full system, but is easier to analyze. The proposed model reduction technique has a simple implementation and may be used for a large class of biochemical reaction models, in the Langevin regime. We tested our method on several realistic models of biochemical kinetics and found an excellent agreement between the dynamics of the full and reduced models.

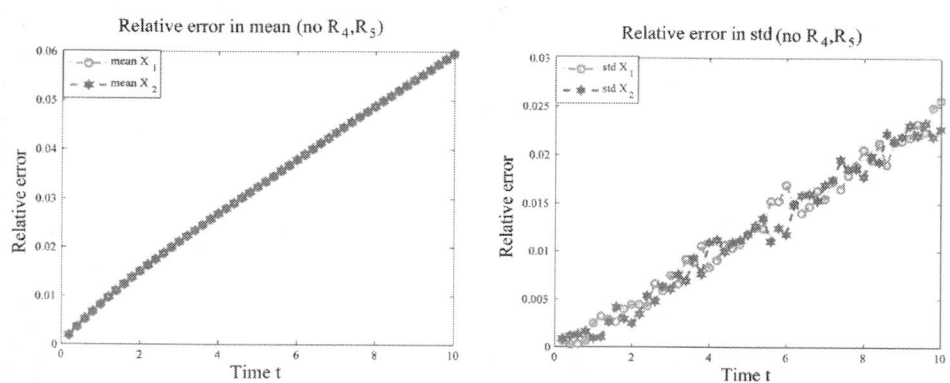

Figure 6: Multiscale biochemical model: the evolution in time of the relative errors in mean (left) and std (right) of X_1 and X_2 in the interval [0, 10], when the reactions R_4 and R_5 are eliminated, compared to the full system. The solution of the Chemical Langevin Equation is approximated over 10,000 trajectories.

ACKNOWLEDGEMENTS

This research was supported by a grant from the Natural Sciences and Engineering Research Council of Canada (NSERC).

REFERENCES

1. H. de Jong, "Modeling and Simulation of Genetic Regulatory Systems," Journal of Computational Biology, Vol. 9, No. 1, 2002, pp. 67-103. doi:10.1089/10665270252833208
2. M. B. Elowitz, A. J. Levine, E. D. Siggia and P. S. Swain, "Stochastic Gene Expression in a Single Cell," Science, Vol. 297, No. 5584, 2002, pp. 1183-1186. doi:10.1126/science.1070919
3. J. M. Raser and E. K. O'Shea, "Control of Stochasticity in Eukaryotic Gene Expression," Science, Vol. 304, No. 5678, 2004, pp. 1811-1814. doi:10.1126/science.1098641
4. D. J. Wilkinson, "Stochastic Modelling for Systems Biology," Chapman & Hall/CRC, Leiden, 2006.
5. S. Ilie and A. Teslya, "An Adaptive Stepsize Method for the Chemical Langevin Equation," The Journal of Chemical Physics, Vol. 136, No. 18, 2012, pp. 184101-184115.doi:10.1063/1.4711143
6. M. S. Okino and M. L. Mavrovouniotis, "Simplification of Mathematical Models of Chemical Reactions," Chemical Reviews, Vol. 98, No. 2, 1998, pp. 391-408. doi:10.1021/cr950223l
7. L. R. Petzold and W. J. Zhu, "Model Reduction for Chemical Kinetics: An Optimization Approach," AICHE Journal, Vol. 44, No. 4, 1999, pp. 869-886. doi:10.1002/aic.690450418
8. H. Huang, M. Fairweather, J. F. Griffiths, A. S. Tomlin and R. B. Brad, "A Systematic Lumping Approach for the Reduction of Comprehensive Kinetic Models," Proceedings of the Combustion Institute, Vol. 30, No.1, 2004, pp. 1309-1316. doi:10.1016/j.proci.2004.08.001
9. G. Li and H. Rabitz, "A General Analysis of Exact Lumping in Chemical Kinetics," Chemical Engineering Science, Vol. 44, No. 6, 1989, pp. 1413-1430.
10. A. Kumar, P. D. Christofides and P. Daoutidis, "Singular Perturbation Modeling of Non-Linear Processes with Non-Explicit Time-Scale Separation," Chemical Engineering Science, Vol. 53, No. 8, 1998, pp. 1491-1504. doi:10.1016/S0009-2509(98)00006-2
11. N. Vora and P. Daoutidis, "Nonlinear Model Reduction of Chemical Reaction System," AICHE Journal, Vol. 47, No. 10, 2001, pp. 2320-2332. doi:10.1002/aic.690471016

12. Y. Cao, D. T. Gillespie and L. Petzold, "The Slow-Scale Stochastic Simulation Algorithm," The Journal of Chemical Physics, Vol. 122, No. 1, 2005, pp. 014116-014134.doi:10.1063/1.1824902
13. C. V. Rao and A. P. Arkin, "Stochastic Chemical Kinetics and the Quasi-Steady-State Assumption: Application to the Gillespie Algorithm," The Journal of Chemical Physics, Vol. 118, No. 11, 2003, pp. 4999-5010. doi:10.1063/1.1545446
14. A. Samant and D. Vlachos, "Overcoming Stiffness in Stochastic Simulation Stemming from Partial Equilibrium: A Multiscale Monte-Carlo Algorithm," The Journal of Chemical Physics, Vol. 123, No. 14, 2005, pp. 144114- 144122. doi:10.1063/1.2046628
15. D. T. Gillespie, "The Chemical Langevin Equations," The Journal of Chemical Physics, Vol. 113, No. 1, 2000, pp. 297-306. doi:10.1063/1.481811
16. D. T. Gillespie, "Exact Stochastic Simulation of Coupled Chemical Reactions," The Journal of Physical Chemistry, Vol. 81, No. 25, 1977, pp. 2340-2361. doi:10.1021/j100540a008
17. D. T. Gillespie, "A General Method for Numerically Simulating the Stochastic Time Evolution of Coupled Chemical Reactions", The Journal of Computational Physics, Vol. 22, No. 4, 1976, pp. 403-434. doi:10.1016/0021-9991(76)90041-3
18. D. T. Gillespie, "A Rigorous Derivation of the Chemical Master Equation," Physica A, Vol. 188, No. 1-3, 1992, pp. 402-425. doi:10.1016/0378-4371(92)90283-V
19. N. G. van Kampen, "Stochastic Processes in Physics and Chemistry," Elsevier, Amsterdam, 2007.
20. A. Varma, M. Morbidelli and H. Wu, "Parametric Sensitivity in Chemical Systems," Cambridge University Press, Cambridge, 1999. doi:10.1017/CBO9780511721779
21. M. Rathinam, P. W. Sheppard and M. Khammash, "Efficient Computation of Parameter Sensitivities of Discrete Stochastic Chemical Reaction Networks," The Journal of Chemical Physics, Vol. 132, No. 3, 2010, pp. 034103- 034116. doi:10.1063/1.3280166
22. P. W. Sheppard, M. Rathinam and M. Khammash, "A Pathwise Derivative Approach to the Computation of Parameter Sensitivities in Discrete Stochastic Chemical Systems," The Journal of Chemical Physics, Vol. 136, No. 3, 2012, pp. 034115-034128.doi:10.1063/1.3677230
23. Y. C. Ho and X. R. Cao, "Optimization and Perturbation Analysis and Queueing Networks," Journal of Optimization Theory and Applications, Vol. 40, No. 4, 1983, pp. 559-582.doi:10.1007/BF00933971
24. R. Suri and M. Zazanis, "Perturbation Analysis Gives Strongly Consistent Sensitivity Estimates for M/G/1 Queue," Management Science, Vol. 34, No. 1, 1988, pp. 39-64.doi:10.1287/mnsc.34.1.39
25. V. Sotiropoulos, M. N. Contou-Carrere, P. Daoutidis and Y. N. Kaznessis, "Model Reduction of Multiscale Chemical Langevin Equations: A Numerical Case Study," IEEE/ACM Transactions on Computational Biology and Bioinformatics, Vol. 6, No. 3, 2009, pp. 470-482. doi:10.1109/TCBB.2009.23

26. T. Jahnke, "On Reduced Models for Chemical Master Equation," Multiscale Modeling and Simulation, Vol. 9, No. 4, 2011, pp. 1646-1676.
27. S. L. Cotter, K. C. Zygalakis, I. G. Kevrekidis and R. Erban, "A Constrained Approach to Multiscale Stochastic Simulation of Chemically Reacting Systems," The Journal of Chemical Physics, Vol. 135, No. 9, 2011, pp. 094102-094114. doi:10.1063/1.3624333.

CITATION

S. Ilie and S. Gholami, "Simplifying Stochastic Mathematical Models of Biochemical Systems," *Applied Mathematics*, Vol. 4 No. 1A, 2013, pp. 248-256. doi: 10.4236/am.2013.41A038.

The Existence and Uniqueness of Random Solution to Itô Stochastic Integral Equation

Hamdin Ahmed Alafif, Caishi Wang
School of Mathematics and Information Science, Northwest Normal University, Lanzhou, China

ABSTRACT

The objective of this paper is to attempt to apply the theoretical techniques of probabilistic functional analysis to answer the question of existence and Uniqueness of a Random Solution to Itô Stochastic Integral Equation. Another type of stochastic integral equation which has been of considerable importance to applied mathematicians and engineers is that involving the Itô or Itô-Doob form of stochastic integrals.

INTRODUCTION

We shall give some historical remarks concerning the development of this type of equation and point out the essential difference between them and other random integral equations.

In 1930 N. Wiener introduced an integral of the form $\int_a^b g(t) dB(\tau)$ where g(t) a deterministic real-valued function and $\{B(\tau), \tau \in [a,b]\}$ is a scalar Brownian motion process.

Author of [1] in 1944 generalized Wiener's integral to include those cases where the integrand is random. That is he obtained an integral of the form

$$\int_0^t g(\tau; w) dB(\tau), t \in [0,1]$$

Which is referred to as the Itô stochastic integral or simply the stochastic integral. Since that time many scientists have contributed to the general development of this type of stochastic integral. For example see [2-10].

In 1946 Author of [5] formulated a stochastic integral equation of the form

$$x(t;w) = C + \int_0^t f(\tau, x(\tau;w)) d\tau$$
$$+ \int_0^t g(\tau, x(\tau;w)) dB(\tau) \qquad (1.0)$$

where $t \in [0,1]$, $\{B(\tau); t \in [0,1]\}$ is a scalar Brownian motion process, and C is a constant Restrictions are usually placed on the functions f and g so that the first integral is interpreted as the usual Lebesgue integral of the sample functions which can then be related to the sample integral of the process $\{f(t, x(t;w)); t \in [0,1]\}$ and the second integral is an Itô stochastic integral.

The principal feature which distinguishes the type of equation studied from an equation of the Itô type is the fact that in the former case each of the integrals involved is interpreted as a Lebesgue integral for almost all $w \in \Omega$. That is, almost all sample functions are Lebesgue integrable. Since in the Itô stochastic integral the limit is taken in the mean-square or in the probability sense, the theory of such integrals has been developed as self-contained and self-consistent.

One of the main purposes of subsequent work in connection with the Itô stochastic integral equation has been to construct Markov processes such that their transition probabilities satisfy given Kolmogorov equations and to investigate the continuity of the processes, among other properties of the sample function.

The method of successive approximation was used by Itô and Doob to show the existence and uniqueness of a random solution to Equation (1.0).

PRELIMINARIES

Let $\{B(t); t \in [a,b]\}$ be a scalar Brownian motion process. In this section we shall be concerned with the integral

$$\int_a^b \phi(t;w)\,dB(t) \quad a < b \tag{1.1}$$

for a fairly general class of functions ϕ. This integral will be called the Itô stochastic integral as we mentioned previously. As is well known, almost all the sample functions of the Brownian motion process are of unbounded variation and hence the integral (1.1) cannot be defined as an ordinary Stieltjes integral.

First we shall define the integral (1.1) for the class of step functions. That is, functions ϕ of the form

$$\phi(t;w) = \begin{cases} 0 & t < a \\ \phi_i(w), & t_i \leq t \leq t_{i+1} \\ 0 & t > b \end{cases} \tag{1.2}$$

where $a = t_0 < t_1 < t_2 < ... < t_{n-1} < t_n = b$, $\phi_i(w)$ are measurable with respect to the σ-algebra A_{t_i}, and

$E\{|\phi_i(w)|^2\} < \infty$ for such functions we define the Itô integral by

$$\int_a^b \phi(t;w)dB(t) = \sum_{i=0}^{n-1} \phi_i(w)(B(t_{i+1}) - B(t_i)) \qquad (1.3)$$

Now suppose that $\phi(t;w)$ is any function satisfying the following conditions.

1) $\phi(t;w)$ is a product-measurable function from $[a,b] \times \Omega \to \mathbb{R}$, assuming the usual Lebesgue measure on $\mathbb{R}_+$.

2) For each, $t \in [a,b]$, $\phi(t;w)$ is measurable with respect to σ-algebra A_t, where A_t is the smallest -algebra on Ω, such that B(s), $s \le t$ is measurable.

3) $\int_{-\infty}^{\infty} E|\phi(t;w)|^2 dt < \infty$

In view of Equation (1.2) it is evident that the class of step functions satisfy conditions 1)-3).

For the function $\phi(t;w)$ satisfying conditions 1)-3) we shall define their norm as follows:

$$\|\phi(t;w)\| = \left\{\int_a^b E\left[|\phi(t;w)|\right]^2 dt\right\}^{\frac{1}{2}} \qquad (1.4)$$

For this case author of [2] has shown the following 1) $\phi(t;w)$ can be approximated in the mean-square sense by a sequence of step functions $\{\phi_n(t;w)\}$. That is

$$\|\phi(t;w) - \phi_n(t;w)\| \to 0 \quad \text{as} \quad n \to \infty$$

2) The sequence of integrals

$$\int_a^b \phi_n(t;w)dB(t)$$

Possesses a mean-square limit. That is there exists a $\theta(w)$ such that

$$E\left\{\theta(w) - \int_a^b \phi_n(t;w)\,dB(t)\right\}^2 \to 0 \tag{1.5}$$

as $n \to \infty$

Now we shall define the integral (1.1) for a class of functions $\{\phi_n(t;w)\}$ satisfying conditions 1)-3) by

$$\int_a^b \phi(t;w)\,dB(t) = \theta(w) \tag{1.6}$$

As with the ordinary integrals, we shall define

$$\int_{-\infty}^{\infty} \phi(t;w)\,dB(t) = \lim_{a\to -\infty, b\to +\infty} \int_a^b \phi(t;w)\,dB(t) \tag{1.7}$$

Definition 1.1 Let $G \in L$, where L denote the collection of Lebesgue measurable subsets of $\mathbb{R}$. Define a function x_G from $\mathbb{R}_+ \times \Omega \to \{0,1\}$ by

$$\chi_G(\tau;w) = \begin{cases} 1 & \text{if } (\tau;w) \in G \times \Omega \\ 0 & \text{otherwise} \end{cases}$$

Lemma 1.1 The function $\varphi X_G : \mathbb{R}_+ \times \Omega \to \mathbb{R}$ defined by

$$(\varphi \chi_G)(\tau;w) = \varphi(\tau;w)\chi_G(\tau;w)$$

where ϕ satisfies conditions 1)-3), and X_G is as defined earlier, also satisfies conditions 1)-3).

Proof. The proof is a straightforward result of the definition of X_G and the fact that ϕ satisfies conditions 1)-3).

We are now in a position to define exactly what is meant by the expression

$$\int_G \phi(\tau;w)\,dB(\tau)$$

Definition 1.2 We define $\int_G \phi(t;w)\,dB(\tau)$ for G a Lebesgue-measurable subset of $\mathbb{R}_+$ by

$$\int_G \varphi(\tau;w)\,dB(\tau) = \int_{-\infty}^{\infty} (\varphi\chi_G)(\tau;w)\,dB(\tau)$$

Note that lemma 1.4 guarantees the expression on the right exists and is well defined **Definition 1.3** We shall denote by

$C^*([a,b], L_2(\Omega,f,P))$ the space of all continuous functions from $[a,b]$ into $L_2(\Omega,f,P)$. We shall define the norm of $C^*([a,b], L_2(\Omega,f,P))$ by

$$\sup_{a \le t \le b} \left\{ \int_\Omega |x(t;w)|^2 \, dP(w) \right\}^{\frac{1}{2}}$$

Lemma 1.2

$$E[\theta] = E\left[\int_{-\infty}^{\infty} \phi(t;w)\,dB(t) \right] = 0$$

Lemma 1.3

$$E\left[|\theta|^2\right] = \int_{-\infty}^{\infty} E\left[|\phi(t;w)|^2\right] dt$$

Lemma 1.4 If we define a distance between two functions ϕ_1 and ϕ_2 each satisfying conditions 1)-3) by

$$\|\phi_1 - \phi_2\| = \left\{ \int_{-\infty}^{\infty} E|\phi_1(t;w) - \phi_2(t;w)|^2 \, dt \right\}^{\frac{1}{2}}$$

and the distance between $\theta_1 = \int_{-\infty}^{\infty} \phi_1(t;w) dB(t)$ and $\theta_2 = \int_{-\infty}^{\infty} \phi_2(t;w) dB(t)$ by

$$\|\theta_1 - \theta_2\| = \left\{ E\left[|\theta_1 - \theta_2|^2 \right] \right\}^{\frac{1}{2}}$$

Then $\|\theta_1 - \theta_2\| = \|\phi_1 - \phi_2\|$.

For the proof of the Lemmas see [2].

Lemma 1.5 Let, $x(t;w) = \int_a^t \phi(t;w) dB(\tau), t \in [a,b]$

Then $x(t;w) \in C^*([a,b], L_2(\Omega, f, P))$

For the proof see [4].

ON AN ITÔ STOCHASTIC INTEGRAL EQUATION

In this section we shall investigate a stochastic integral equation of the type

$$x(t;w) = \int_0^t k(t,\tau;w) f(\tau, x(\tau;w)) d\tau$$
$$+ \int_0^t \phi(\tau;w) dB(\tau) \quad t \geq 0 \qquad (2.1)$$

where $x(t;w)$ is the unknown random process defined for $t \in \mathbb{R}$ and $w \in \Omega$.

We shall place the following restrictions on the random functions which constitute the stochastic integral Equation (2.1).

1') $k(t,r;w)$ is an element of $L_\infty(\Omega, f, P)$ and $k(t,r;w): \Delta \to L_\infty(\Omega, f, P)$ is continuous where $\Delta\{(t,\tau): 0 \leq \tau \leq t < \infty\}$.

2') $x(t;w) \to f(t,x(t;w))$ is an operator on the set S with values in the Banach space B satisfying

$$\|f(t,x(t;w)) - f(t,y(t;w))\|_B$$
$$\leq \lambda \|x(t;w) - y(t;w)\|_D$$

For $x(t;w), y(t;w) \in S$.

3') Conditions 1)-3) of section 1 hold.

Thus with the given assumptions the first integral of (2.1) can be interpreted as a Lebesgue integral and the second as an Itô stochastic integral.

We shall now proceed to state and prove a theorem concerning the behavior of the Itô integral. More precisely, if we show that the Itô integral is an element of the space $C_c(\mathbb{R}_+, L_2(\Omega, f, P))$, we can apply the theory of admissibility to Equation (2.1) to show the existence of a random solution. By a random solution to Equation (2.1) we mean a random function $x(t;w)$ from $\mathbb{R}_+$ into $L_2(\Omega, f, P)$ such that for each $t \in \mathbb{R}_+$, $x(t;w)$ satisfies the integral equation P-a.e. showing that the Itô integral is an element of $C_c(\mathbb{R}_+, L_2(\Omega, f, P))$ will make feasible the assumption that we wish to make that the integral is an element of D, a Banach space contained in the topological space mentioned For convenient we shall denote the Itô integral by

$$h(t;w) = \int_0^t \phi(\tau;w) \, dB(\tau), \quad t \geq 0$$

The Existence and Uniqueness of Random Solution to Itô Stochastic

Theorem 2.1 For

$t \in \mathbb{R}_+, h(t;w) \in C_c\left(\mathbb{R}_+, L_2(\Omega, \mathcal{F}, P)\right)$

Proof Fix $t \in \mathbb{R}_+$ Then

$$h(t;w) = \int_0^t \phi(\tau;w) dB(\tau) = \int_{-\infty}^{\infty} \left(\phi x_{[0,t]}\right)(\tau;w) dB(\tau)$$

Thus

$$E\left|h(t;w)\right|^2 = E\left|\int_{-\infty}^{\infty} \left(\phi x_{[0,t]}\right)(\tau;w) dB(\tau)\right|^2$$

$$= \int_{-\infty}^{\infty} E\left|\left(\phi x_{[0,t]}\right)(\tau;w)\right|^2 d\tau$$

by lemma 1.3.

Hence $E|h(t;w)|^2 < \infty$.

Therefore for fixed t, $h(t;w) \in L_2(\Omega,f,P)$. Now let $t_n \to t$ in $\mathbb{R}_+$. To show that $h_n(t;w) \to h(t;w)$ in $L_2(\Omega,f,P)$, it is sufficient to show that

$$\left\|\phi x_{[0,t_n]} - \phi x_{[0,t]}\right\|$$

can be made arbitrarily small. That is, we must show that

$$\int_{-\infty}^{\infty} E\left|\left(\phi x_{[0,t_n]}\right)(\tau;w) - \left(\phi x_{[0,t]}\right)(\tau;w)\right|^2 dt$$

Can be made arbitrarily small. Choose $\varepsilon > 0$. Consider the nonnegative function $q(\tau;w) \to E|(\tau;w)|^2$. By condition 3) $q(\tau;w)$ is integrable over $\mathbb{R}_+$. Hence there exists a $\delta > 0$ such that for every set of Lebesgue measure less than δ, $\int_G q(\tau;w) d\tau < \infty$. Thus

$$\int_{-\infty}^{\infty} E \left| \left(\phi \chi_{[0,t_n]} \right)(\tau;w) - \left(\phi \chi_{[0,t]} \right)(\tau;w) \right|^2 dt$$

$$= \int_{t}^{t_n} E \left| \left(\phi \chi_{[0,t_n]} \right)(\tau;w) - \left(\phi \chi_{[0,t]} \right)(\tau;w) \right|^2 d\tau$$

$$= \int_{t}^{t_n} E |\phi(\tau;w)|^2 d\tau = \int_{t}^{t_n} q(\tau;w) d\tau$$

Since for $n > N_\delta$ and $|t_n - t| < \delta$ and since the Lebesgue measure of the interval (t, t_n) is its length, we conclude that the Lebesgue measure of (t, t_n) is less than δ.

Hence

$$\int_{t}^{t_n} q(\tau;w) d\tau < \varepsilon$$

Implying that $t \to h(t;w)$ is continuous from $\mathbb{R}_+$ into $L_2(\Omega, f, P)$ and the proof is complete.

Since we have shown that $h(t;w) \in C_c(\mathbb{R}_+, L_2(\Omega, f, P))$, we can conclude that the stochastic integral Equation (2.1) possesses a unique random solution

ON ITÔ-DOOB-TYPE STOCHASTIC INTEGRAL EQUATIONS

In this section we shall study the existence and uniqueness of a random solution to a stochastic integral equation of the form

$$x(t;w) = \int_{0}^{t} f(\tau, x(\tau;w)) d\tau + \int_{0}^{t} \phi(\tau, x(\tau;w)) dB(\tau) \qquad (3.1)$$

Where $t \in [0,1]$. As before, the first integral is a Lebesgue integral, while the second is an Itô-type stochastic integral defined with respect to a scalar Brownian motion process $\{B(t), t \in [0,1]\}$.

The Existence and Uniqueness of Random Solution to Itô Stochastic

Recall that
$C^*([0,1], L_2(\Omega, f, P)) \subset C_c(\mathbb{R}_+, L_2(\Omega, f, P))$, We shall define the operators W_1 and W_2 from $C^*([0,1], L_2(\Omega, f, P))$ into $C^*([0,1], L_2(\Omega, f, P))$ by

$$(W_1 x)(t; w) = \int_0^t x(\tau; w) d\tau \tag{3.2}$$

and

$$(W_2 x)(t; w) = \int_0^t x(\tau; w) dB(\tau) \tag{3.3}$$

Note that in view of lemma 1.5 $x(t; w) \in C^*([0,1], L_2(\Omega, f, P))$. Its clear that W_1 and W_2 are linear operators.

Theorem 3.1 The operators W_1 and W_2 defined by (3.2) and (3.3) respectively, are continuous operators from $C^*([0,1], L_2(\Omega, f, P))$ into $C^*([0,1], L_2(\Omega, f, P))$.

Lemma 3.1 Let T be a continuous operator from $C_c(\mathbb{R}, L_2(\Omega, f, P))$ into itself. If B and D are Banach spaces stronger than C_c and the pair (B, D) is admissible with respect to T. Then T is a continuous operator from B to D.

Proof of theorem 3.1 The fact that W_1 is a continuous operator from $C^*([0,1], L_2(\Omega, f, P))$ into $C^*([0,1], L_2(\Omega, f, P))$ follows from lemma 3.1. From (3.3) we have

$$\|(W_2 x)(t; w)\|_{L_2(\Omega, \mathcal{F}, P)}^2 = \int_\Omega dP(w) \left\{ \int_0^t x(\tau; w) dB(\tau) \right\}^2$$
$$= \int_0^t d\tau \int_\Omega x^2(\tau; w) dP(w)$$

Furthermore

$$\|(w_2x)(t;w)\|^2_{L_2(\Omega,\mathcal{F},P)} \leq \int_0^t d\tau \sup_{0\leq\tau\leq t} \|x(\tau;w)\|^2_{L_2(\Omega,\mathcal{F},P)}$$

$$= \int_0^t \left\{\sup_{0\leq\tau\leq t} \|x(\tau;w)\|_{L_2(\Omega,\mathcal{F},P)}\right\}^2 d\tau$$

$$= \|x(t;w)\|^2 \int_0^t d\tau \leq \|x(t;w)\|^2$$

Therefore

$$\|(w_2x)(t;w)\| \leq \|x(t;w)\|$$

Thus W_1 and W_2 are continuous operators from $C^*([0,1],L_2(\Omega,f,P))$ into $C^*([0,1],L_2(\Omega,f,P))$.

An Existence Theorem

We shall assume that lemma 3.1 holds with respect to the operators W_1 and W_2. Therefore there exist positive constants K_1 and K_2 less than one such that

$$\|(w_1x)(t;w)\|_D \leq k_1 \|x(t;w)\|_B$$

And

$$\|(w_2x)(t;w)\|_D \leq k_2 \|x(t;w)\|_B$$

The following theorem gives sufficient conditions for the existence of a unique random solution, a second order stochastic process, to the Itô-Doob stochastic integral Equation (3.1).

Theorem 3.2 Consider the stochastic integral equation (3.1) under the following condition:

1) B and D are Banach spaces in $C^*([0,1], L_2(\Omega, f, P))$ which are stronger than $C^*([0,1], L_2(\Omega, f, P))$ such that (B,D) is admissible with respect to the operators W_1 and W_2.

2) a) $x(t;w) \to f(t, x(t;w))$ is an operator on
$$S = \{x(t;w) : x(t;w) \in D \text{ and } \|x(t;w)\|_D \leq \rho\}$$

With values in B satisfying
$$\|f(t, x(t;w)) - f(t, y(t;w))\|_B \leq \lambda_1 \|x(t;w) - y(t;w)\|_D$$

b) $x(t;w) \to \phi(t, x(t;w))$ is an operator on S into B satisfying
$$\|\phi(t, x(t;w)) - \phi(t, y(t;w))\|_B \leq \lambda_2 \|x(t;w) - y(t;w)\|_D$$

where λ_1 and λ_2 are constants. Then there exists a unique random solution to Equation (3.1) provided that $k_1 \lambda_1 + k_2 \lambda_2 < 1$. And
$$\|f(t, 0)\|_B + \|\phi(t, 0)\|_B \leq \rho(1 - \lambda_1 k_1 - \lambda_2 k_2)$$

Proof. Define an operator U from the set S into D as follows
$$(Ux)(t;w) = \int_0^t f(\tau, x(\tau; w)) d\tau + \int_0^t \phi(\tau, x(\tau; w)) dB(\tau)$$

We need to show that U is a contraction operator on S and that $US < S$.

Let $x(t;w), y(t;w) \in S$.

Then $(Ux)(t;w) - (Uy)(t;w) \in D$ because D is a Banach space. Further, we have

$$\|(Ux)(t;w) - (Uy)(t;w)\|_D$$
$$\leq \left\| \int_0^t [f(\tau, x(\tau;w)) - f(\tau, y(\tau;w))] d\tau \right\|_D$$
$$+ \left\| \int_0^t [\phi(\tau, x(\tau;w)) - \phi(\tau, y(\tau;w))] dB(\tau) \right\|_D$$
$$\leq k_1 \|f(t, x(t;w)) - f(t, y(t;w))\|_B$$
$$+ k_2 \|\phi(t, x(t;w)) - \phi(t, y(t;w))\|_B$$
$$\leq (\lambda_1 k_1 + \lambda_2 k_2) \|x(t;w) - y(t;w)\|_D < \|x(t;w) - y(t;w)\|$$

Thus U is a contraction operator.

For any element in S we have

$$\|(Ux)(t;w)\|_D$$
$$\leq \left\| \int_0^t f(\tau, x(\tau;w)) d\tau \right\|_D + \left\| \int_0^t \phi(\tau, x(\tau;w)) dB(\tau) \right\|_D$$
$$\leq k_1 \|f(t, x(t;w))\|_B + k_2 \|\phi(t, x(t;w))\|_B$$
$$\leq \lambda_1 k_1 \|x(t;w)\|_D + \lambda_2 k_2 \|x(t;w)\|_D$$
$$+ k_1 \|f(t,0)\|_B + k_2 \|\phi(t,0)\|_B$$

Since $x(t;w) \in s$ it follows that

$$\|(Ux)(t;w)\|_D$$
$$\leq \rho(\lambda_1 k_1 + \lambda_2 k_2) + \|f(t,0)\|_B + \|\phi(t,0)\|_B \leq \rho$$

from the assumptions in the theorem.

Thus the existence and uniqueness of a random solution to Equation (3.1) follow from the Banach fixed-point theorem.

Theorem 3.4 (S. Banach's fixed-point principle) ([11]).

If T is a contraction operator on a complete metric space H. then there exists a unique point $x^* \in H$ for which $T(x^*) = x^*$.

CONCLUSION

We investigated the existence and uniqueness of Itô stochastic integral equation by applying the theoretical techniques of probabilistic functional analysis. In fact author of [12] refers to probabilistic functional analysis as being concerned with the applications and extensions of the methods of functional analysis to the study of the various concepts, processes, and structures which arise in the theory of probability and its applications. Finally to develop and unify the theory of stochastic or random equations see [13-15].

REFERENCES

1. K. Ito, "Stochastic Integral," Proceedings of the Imperial Academy, Vol. 20, No. 8, 1944, pp. 519-524. doi:10.3792/pia/1195572786
2. J. L. Doob, "Stochastic Processes," Wiley, New York, 1953, pp. 426-432.
3. Y. Dynkin, "Markov Processes," Academic Press, New York, 1964, pp. 9-13.
4. A. Jazwinski, "Stochastic Processes and Filtering Theory. Mathematics in Science and Engineering," Vol. 64, Academic Press, New York, 1970, pp. 97-105
5. K. Ito, "On a Stochastic Integral Equation," Proceedings of the Japan Academy, Vol. 22, No. 2, 1946, pp. 32-35. doi:10.3792/pja/1195572371
6. H. P. Mckean, "Stochastic Integrals," Academic Press, New York, 1969, pp. 21-25.

7. T. L. Satty, "Modern Nonlinear Equations," McGrowHill, New York, 1967, pp. 216-226.
8. L. Gikhmann and A. V. Skorokhod, "Introduction to the Theory of Random Process-Saunders," Philadehphia, Pennsylvania, 1969, pp. 378-391.
9. R. L. Stratonovich, "A New Representation for Stochastic Integrals and Equations," Journal of SLAM Control, Vol. 4, 1966, pp. 362-371.
10. E. Wong and M. Zakai, "On the Relation between Ordinary and Stochastic Differential Equations," International Journal of Engineering Science, Vol. 3, No. 2, 1965, pp. 213-229. doi:10.1016/0020-7225(65)90045-5
11. I. P. Natanson, "Theory of Functions of a Real Variable," Vol. II, Ungar, New York, 2010.
12. A. T. Bharucha-Reid, "On the Theory of Random Equations," Proceedings of Symposia in Applied Mathematics, Vol. 16, 1964, pp. 40-69.
13. G. Adomain, "Random Operator Equations in Mathematical Physics," Journal of Mathematical Physics, Vol. 11, No. 3, 1970, pp. 1069-1074. doi:10.1063/1.1665198
14. G. Adomain, "Linear Random Operator Equations in Mathematical Physics III," Journal of Mathematical Physics, Vol. 12, No. 9, 1971, pp. 1944-1948. doi:10.1063/1.1665827
15. G. Adomain, "Theory of Random Systems," Transactions of the fourth Prague Conference on Information Theory, Statistical Decision Functions, Random Processes, Prague, 31 August-11 September 1965, pp. 205-222.

CITATION

H. Alafif and C. Wang, "The Existence and Uniqueness of Random Solution to Itô Stochastic Integral Equation," *Applied Mathematics*, Vol. 3 No. 7, 2012, pp. 800-804. doi: 10.4236/am.2012.37119.

Solving Nonlinear Stochastic Diffusion Models With Nonlinear Losses Using The Homotopy Analysis Method

Aisha A. Fareed[1], Hanafy H. El-Zoheiry[2], Magdy A. El-Tawil[2], Mohammed A. El-Beltagy[3], Hany N. Hassan[1]

[1]Department of Basic Sciences, Engineering Faculty, Benha University, Benha, Egypt
[2]Department of Engineering Mathematics & Physics, Engineering Faculty, Cairo University, Cairo, Egypt
[3]Department of Electrical & Computer Engineering, Engineering Faculty, Effat University, Jeddah, KSA

ABSTRACT

This paper deals with the construction of approximate series solutions of diffusion models with stochastic excitation and nonlinear losses using the homotopy analysis method (HAM). The mean, variance and other statistical properties of the stochastic solution are computed. The solution technique was applied successfully to the 1D and 2D diffusion models. The scheme shows importance of choice of convergence-control parameter ⊠ to guarantee the convergence of the solutions of nonlinear differential Equations. The results are compared with the Wiener-Hermite expansion with perturbation (WHEP) technique and good agreements are obtained.

INTRODUCTION

The deterministic differential equations of the form $\dot{x}(t) = a(t)x(t)$ constitute the basic form of so-called diffusion or transport problems which appear in relevant models such as: the growth population geometric (or Malthusian) model in biology, where $a(t)$ represents the per capita growth rate; the neutron and gamma ray transport model in physics, where coefficient a(t) involves the geometry of the cross-sections of the medium; the continuous composed interest rate models for studying the evolution of an investment under time-variable interest rate r(t) which can be taken as a(t)=1+r(t), etc. Despite the usefulness of these basic models, they do not often cover all possible situations observed from a practical point of view. In fact, as a simple but illustrative example, if a(t)=a>0, the Malthus model predicts unlimited growth of a species despite the fact that resources are always limited. Then, the logistic (or Verhulst) model introduces a nonlinear term in order to overcome this drawback by considering the differential equation $\dot{x}(t) = a(t)x(t) - bx(t)^2, a, b > 0$, where the nonlinearity intensity is given by parameter b. In many practical situations it is appropriate to assume that the nonlinear term affecting the phenomena under study is small enough; then its intensity is controlled by means of a frank small parameter, say ε. Stochastic differential equations based on the white noise process provide a powerful tool for dynamically modeling these complex and uncertain aspects. Over the last few years, new and relevant methods for finding the exact solutions of such Equations have been developed. They include the homotopy perturbation (HPM) method [1,2], Wiener-Hermite expansion with perturbation method (WHEP Cortes [2011]) [3] and the exp-function method [4,5].

HAM is an analytical technique for solving non linear differential equations. Proposed by Liao in 1992, [6], the technique is superior to the traditional perturbation methods, in which it leads to convergent series solutions of strongly nonlinear problems, independent of any small or large physical parameter associated with the problem, [7]. The HAM provides a more viable alternative to non perturbation techniques such as the Adomian decomposition method (ADM) [8] and other techniques that cannot guarantee the convergence of the solution series and may be only valid for weakly nonlinear problems, [7]. We note here that He's HPM method, [9] is only a special case of the HAM. In recent years, this method has been successfully employed to solve many problems in science and engineering such as the viscous flows of non-Newtonian fluids [10,11], the KdV-type equations [12],

Glauert-jet problem [13], Burgers-Huxley equation [14], time-dependent Emden-Fowler type equations [15], differential-difference equation [16], two-point nonlinear boundary value problems [17]. The HAM provides the solution in the form of a rapidly convergent series with easily computable components using symbolic computation software such as Mathematica.

This paper deals with the solution of 1D stochastic differential models of the form

$$\dot{x}(t) = a(t)x(t) - \varepsilon x^2(t) + \lambda n(t;\omega), \quad t > 0,$$
$$x(0) = x_0 \quad (1)$$

where the diffusion coefficient a (t) and initial condition x_0 are deterministic, ε is a small parameter and n(t;ω) is the white noise process, whose intensity is given by parameter λ, which has the following important properties:

$$E[n(t;\omega)] = 0$$
$$E[n(t_1;\omega)n(t_2;\omega)] = \delta(t_1 - t_2)$$

where E denotes the ensemble average operator, δ is the Dirac delta function. And ω is a random outcome for a triple probability space (Ω, A, P), where Ω is a sample space, A is a σ-algebra associated with Ω and P is a probability measure. The current work also deals with the solution of 2D stochastic quadratic nonlinear equation with σn(x;ω) as non-homogeneity.

$$\frac{\partial u(t,x;\omega)}{\partial t} = \frac{\partial^2 u}{\partial x^2} - \varepsilon u^2 + \sigma n(x;\omega)$$

$$(t,x) \in (0,\infty) \times (0,l),$$

$$u(t,0) = 0, u(t,l) = 0 \text{ and } u(0,x) = \phi(x). \tag{2}$$

Where u(t,x;ω) is the diffusion process, ε is a deterministic scale for the nonlinear term. The non-homogeneity term σn(x;ω) is spatial white noise scaled by σ.

The paper is organized as follows. Section 2 summarizes the basic idea of the HAM method. In Section 3, the HAM is applied in order to obtain fourth order approximation of the solution of 1D diffusion model. In Section 4, the HAM is applied up to the third order approximation for the solution of 2D diffusion model. In addition, we compute approximations for the main statistical moments such as the mean and variance. A comparison is done with the results obtained with the (WHEP Cortes [2011], WHEP El-Beltagy [2013]) technique [4,5]. The results are shown in Section 5 along with comments on the results.

THE BASIC IDEA OF HAM

A presentation of the standard HAM for deterministic problems can be found in [9]. The following subsection is a brief description of HAM. Consider the following differential equation:

$$N[u(t,x)] = 0 \tag{3}$$

where N is a nonlinear operator and u(t,x) is the unknown function. By means of generalizing the traditional HPM method, Liao [6] constructs the so-called zero-order deformation equation

$$(1-q)L[\phi(t,x;q) - u_0(t,x)] = q\hbar H(t,x) N[\phi(t,x;q)], \tag{4}$$

Where $q \in [0,1]$ denotes the so-called embedding parameter, $\hbar \neq 0$ is an auxiliary parameter and L is an auxiliary linear operator.

Solving Nonlinear Stochastic Diffusion Models With Nonlinear Losses

The HAM is based on a kind of continuous mapping $u(t,x) \rightarrow \phi(t,x;q)$, where $\phi(t,x;q)$ is an unknown function, $u_0(t,x)$ is an initial guess for $u(t,x)$, and $H(t,x)$ denotes a non-zero auxiliary function. It is obvious that when the embedding parameter q=0 and q=1, Equation (3) becomes

$$\phi(t,x;0) = u_0(t,x), \quad \phi(t,x;1) = u(t,x), \tag{5}$$

respectively. Thus as q increases from 0 to 1, the solution $\phi(t,x;q)$ varies from the initial guess $u_0(t,x)$ to the solution $u(t,x)$. In topology, this kind of variation is called deformation; Equation (3) constructs the homotopy $\phi(t,x;q)$.

Having the freedom to choose the auxiliary parameter $\hbar$, the auxiliary function $H(t,x)$, the initial approximation $u_0(t,x)$, and the auxiliary linear operator L, we can assume that all of them are properly chosen so that the solution $\phi(t,x;q)$ of the zero-order deformation Equation (4) exists for $0 \leq q \leq 1$.

Expanding $\phi(t,x;q)$ in Taylor series with respect to q, one has,

$$\phi(t,x;q) = u_0(t,x) + \sum_{m=1}^{\infty} u_m(t,x) q^m, \tag{6}$$

where

$$u_m(t,x) = \frac{1}{m!} \frac{\partial^m \phi(t,x;q)}{\partial q^m} \bigg|_{q=0} \tag{7}$$

Assume that the auxiliary parameter $\hbar$, the auxiliary function $H(t,x)$, the initial approximation $u_0(t,x)$ and the auxiliary linear operator L are so properly chosen that the series (6) converges at q=1 and

$$\phi(t,x;1) = u_0(t,x) + \sum_{m=1}^{\infty} u_m(t,x), \qquad (8)$$

which must be one of the solutions of the original nonlinear Equation, as proved by Liao [9]. As $\hbar = -1$ and $H(t,x)=1$ Equation (4) becomes

$$(1-q)L\big[\phi(t,x;q) - u_0(t,x)\big] + qN\big[\phi(t,x;q)\big] = 0, \qquad (9)$$

This is mostly used in the HPM method. According to definition (8), the governing equation and the corresponding initial condition of $u_m(t,x)$ can be deduced from the zero-order deformation Equation (4). Define the vector

$$\boldsymbol{u}_n(t,x) = \{u_0(t,x), u_1(t,x), u_2(t,x), \cdots, u_n(t,x)\}.$$

Differentiating Equation (4) m times with respect to the embedding parameter q and then setting q=0 and finally dividing them by m!, we have the so-called m^{th}-order deformation equation:

$$L\big[u_m(t,x) - \chi_m u_{m-1}(t,x)\big] = \hbar H(t,x) R(u_{m-1}); \; m \geq 1 \qquad (10)$$

where

$$R(u_{m-1}) = \frac{1}{(m-1)!} \frac{\partial^{m-1} N\big[\phi(t,x;q)\big]}{\partial q^{m-1}}\bigg|_{q=0},$$

and

$$\chi_m = \begin{cases} 0 & \text{when } m \leq 1 \\ 1 & \text{otherwise} \end{cases}, \qquad (11)$$

Solving Nonlinear Stochastic Diffusion Models With Nonlinear Losses

The solution is computed as:

$$u(t,x) = \sum_{i=0}^{\infty} u_i(t,x).$$

It should be emphasized that $u_m(t,x)$ for $m \geq 1$ is governed by the linear Equation (10) with linear boundary conditions that come from the deterministic problem, which can be solved by any symbolic computation software such as Mathematica, Maple, or Matlab.

APPLICATION TO THE 1D DIFFUSION MODEL

To demonstrate the above presented method it will be used to find the mean and variance of 1D stochastic diffusion problem as follows.

The auxiliary linear operator will be chosen as

$$L[\phi(t;q)] = \frac{d\phi(t;q)}{dt}$$

Furthermore, we define the nonlinear operator as

$$N[\phi(t;q)] = \frac{d\phi(t;q)}{dt} - a(t)\phi(t;q) + \varepsilon(\phi(t;q))^2 + \lambda n(t);$$

We construct the zero-order deformation equation,

$$(1-q)L[X_m(t) - \chi_m X_{m-1}(t)] = q\hbar H(t) R(X_{m-1}).$$

The mth-order deformation equation for $m \geq 1$ and $H(t)=1$ is

$$L[X_m(t) - \chi_m X_{m-1}(t)] = \hbar R(X_{m-1}),$$

(12)

Subject to the initial condition

$$X_m(0) = 0,$$

where

$$R(X_{m-1}) = \frac{dX_{m-1}(t)}{dt} - a(t)X(t) + \varepsilon \sum_{i=0}^{m-1} X_{m-1-i}(t)X_i(t) - (1-\chi_m)\lambda n(t);$$

Now the solution of the mth-order deformation Equation (12) for m ≥ 1 becomes

$$L[X_m(t) - \chi_m X_{m-1}(t)] = \hbar \left[\frac{dX_{m-1}(t,x)}{dt} - a(t)X(t) + \varepsilon \sum_{i=0}^{m-1} X_{m-1-i}(t)X_i(t) - (1-\chi_m)\lambda n(t) \right]$$

The first order approximation is obtained by setting m=1 in (12) as follows

$$L[X_1(t)] = \hbar R(X_0)$$

where

$$R(X_0) = \frac{dX_0(t)}{dt} - a(t)X(t) + \varepsilon X_0^2 - \lambda n(t)$$

Then

$$L[X_1(t)] = \hbar \left[\frac{dX_0(t)}{dt} - a(t)X_0(t) + \varepsilon X_0^2(t) - \lambda n(t) \right],$$

$$X_1(t) = \hbar \int_0^t \left[\frac{dX_0(t)}{dt} - a(t)X_0(t) + \varepsilon X_0^2(t) - \lambda n(t) \right] dt,$$

Solving Nonlinear Stochastic Diffusion Models With Nonlinear Losses

The ensemble average of the first order approximation is

$$E[X_1(t)] = \hbar \int_0^t \left[\frac{dE[X_0(t)]}{dt} - a(t)E[X_0(t)] + \varepsilon E[X_0^2(t)] \right] dt,$$

$$E[X_1(t)] = -0.2475\hbar t$$

The covariance of the first order solution will be

$$\text{Cov}[X_1(t_1), X_1(t_2)] = E[(X_1(t_1) - EX_1(t_1))(X_1(t_2) - EX_1(t_2))]$$

$$= E\left[\left(-\hbar\lambda \int_0^{t_1} n(t_1) dt_1\right)\left(-\hbar\lambda \int_0^{t_2} n(t_2) dt_2\right)\right] = \hbar^2\lambda^2 \int_0^{t_2}\int_0^{t_1} E[n(t_1)\cdot n(t_2)] dt_1 dt_2$$

$$= \hbar^2\lambda^2 \int_0^{t_2}\int_0^{t_1} \delta(t_1 - t_2) dt_1 dt_2 \quad = \hbar^2\lambda^2 \int_0^{t_1} dt_1 = \hbar^2\lambda^2 t_1$$

The variance of the first order solution will be

$$\text{Var}[X_1(t)] = E[X_1(t) - EX_1(t)]^2 = E\left[-\hbar\lambda \int_0^t n(t) dt\right]^2 = \hbar^2\lambda^2 t$$

In this manner, we can have more results of E [$X_m(t)$] and Var[$X_m(t)$] obtained at m=2,3,4,...

The final expression of the mean of the 4th order solution will be

$$E[X(t)] = E\left[\sum_{i=0}^{M=4} X_i(t)\right] = 0.5 - 0.99\hbar t + \hbar^2(-1.4849 + 0.3638t)t$$

$$+ \hbar^3 t(-0.99 + 0.4951t - 0.0394123125t^2) + \hbar^4 t(-0.2475 + 0.1869t - 0.0298t^2 + 0.0012t^3)$$

Since $X(t) = \sum_{i=1}^{N} X_i(t)$

Then the final expression of the variance of the 2nd order solution will be

$$\text{Var}\left(\sum_{i=1}^{N} X_i(t)\right) = \sum_{i=1}^{N} \text{Var}[X_i(t)] + \left(\sum_{i=1}^{N}\sum_{j\neq i}^{N} \text{Cov}[X_i(t), X_j(t)]\right) \text{Var}[X(t)]$$

$$= \text{Var}[X_1(t)] + \text{Var}[X_2(t)] + 2\text{Cov}[X_1(t_1), X_2(t_2)] \text{Var}[X(t)]$$

$$= h^2 t \left(4. + h(4. - 1.7149t) + h^2\left(1. - 0.735t + 0.12t^2\right)\right)$$

APPLICATION TO THE 2D DIFFUSION MODEL

HAM will be used to find mean and variance of stochastic quadratic nonlinear diffusion problem as follows.

The auxiliary linear operator is chosen as

$$L[\phi(t,x;q)] = \frac{\partial \phi(t,x;q)}{\partial t} - \frac{\partial^2 \phi(t,x;q)}{\partial x^2}$$

We have many choices in guessing the initial approximation together with its initial conditions which greatly affects the consequent approximation. The choice u_0 is a design problem which can be taken as follows:

$$u_0(t,x) = \sum_{n=0}^{\infty} B_n e^{\beta_n t} \sin\frac{n\pi}{\ell}x \qquad (13)$$

$$B_n = \frac{2}{\ell} \int_0^{\ell} \phi(x) \sin\frac{n\pi}{\ell} x \, dx$$

Solving Nonlinear Stochastic Diffusion Models With Nonlinear Losses

One can notice that the selected value function satisfies the initial and boundary conditions and it depends on the parameter β_n which is totally free. One can also notice that β_n selection could control the solution convergence.

Furthermore, we define the nonlinear operator as

$$N[\phi(t,x;q)] = \frac{\partial \phi(t,x;q)}{\partial t} - \frac{\partial^2 \phi(t,x;q)}{\partial x^2} + \varepsilon[\phi(t,x;q)]^2 - \sigma n(x;\omega)$$

We construct the zero-order deformation Equation,

$$(1-q)L[u_m(t,x) - \chi_m u_{m-1}(t,x)] = q\hbar H(t,x) R(u_{m-1}).$$

The mth-order deformation Equation for $m \geq 1$ and $H(t,x)=1$ is

$$L[u_m(t,x) - \chi_m u_{m-1}(t,x)] = \hbar R(u_{m-1}),$$

And subject to the boundary conditions

$$u_m(t,0) = 0, u_m(t,l) = 0$$

And the initial condition

$$u_m(0,x) = 0,$$

where

$$R(u_{m-1}) = \frac{\partial u_{m-1}(t,x)}{\partial t} - \frac{\partial^2 u_{m-1}(t,x)}{\partial x^2} + \varepsilon \left[\sum_{i=0}^{m-1} u_{m-1-i}(t,x) u_i(t,x)\right] - (1-\chi_m)\sigma n(x;\omega).$$

Now the mth-order deformation equation for $m \geq 1$ becomes

$$L\left[u_m(t,x)-\chi_m u_{m-1}(t,x)\right]=\hbar\left[\frac{\partial u_{m-1}(t,x)}{\partial t}-\frac{\partial^2 u_{m-1}(t,x)}{\partial x^2}\right]+$$

$$\hbar\varepsilon\sum_{i=0}^{m-1}u_{m-1-i}(t,x)u_i(t,x)-\hbar(1-\chi_m)\sigma n(x;\omega).$$

The first order approximation is obtained by substituting m=1 to get

$$L\left[u_1(t,x)\right]=\hbar\left[\frac{\partial u_0(t,x)}{\partial t}-\frac{\partial^2 u_0(t,x)}{\partial x^2}+\varepsilon u_0^2-\sigma n(x;\omega)\right]. \tag{14}$$

The approximated first order solution of (14) can be obtained using Eigen function expansion as follows,

$$u_1(t,x)=\sum_{n=0}^{\infty}I_{n,1}(t)\sin\frac{n\pi}{\ell}x$$

Where

$$I_{n,1}(t)=\int_0^t e^{\left(\frac{-n\pi}{\ell}\right)^2(t-\tau)}F_{n,1}(\tau)d\tau$$

$$F_{n,1}(t)=\frac{2\hbar}{\ell}\int_0^\ell\left[\frac{\partial u_0(t,x)}{\partial t}-\frac{\partial^2 u_0(t,x)}{\partial x^2}+\varepsilon u_0^2-\sigma n(x;\omega)\right]\sin\frac{n\pi}{\ell}x\,dx,$$

the ensemble average of the first order approximation is

$$\mu\left[u_1(t,x)\right]=\sum_{n=0}^{\infty}E\left(I_{n,1}(t)\right)\sin\frac{n\pi}{\ell}x$$

Solving Nonlinear Stochastic Diffusion Models With Nonlinear Losses

Where

$$E(I_{n,1}(t)) = \int_0^t e^{\left(\frac{-n\pi}{\ell}\right)^2 (t-\tau)} E(F_{n,1}(\tau)) d\tau$$

$$E(F_{n,1}(t)) = \frac{2\hbar}{\ell} \int_0^L \left[\frac{\partial u_0(t,x)}{\partial t} - \frac{\partial^2 u_0(t,x)}{\partial x^2} + \varepsilon u_0^2\right] \sin\frac{n\pi}{\ell} x dx \mu [u_1(t,x)]$$

$$= \frac{e^{-\pi^2 t} \hbar \left(3\left(-1+e^{t(\pi^2+\beta n)}\right)\pi(\pi^2+2\beta n) - 8\varepsilon + 8e^{t(\pi^2+2\beta n)}\varepsilon\right)\sin[\pi x]}{3(\pi^3 + 2\pi\beta n)}.$$

The covariance of the first order solution can be computed as

$$\text{Cov}[u_1(t,x_1), u_1(t,x_2)] = E\left[(u_1(t,x_1) - Eu_1(t,x_1))(u_1(t,x_2) - Eu_1(t,x_2))\right]$$

$$= E\left[\left(\sum_{n=1}^{\infty}(I_{n,1}(t) - EI_{n,1}(t))\sin\frac{n\pi}{\ell} x_1\right)\left(\sum_{m=1}^{\infty}(I_{m,1}(t) - EI_{m,1}(t))\sin\frac{m\pi}{\ell} x_2\right)\right].$$

The covariance is obtained from the following final expression

$$\text{Cov}(u_1(t,x_1), u_1(t,x_2))$$

$$\frac{4\hbar^2 \sigma^2}{\ell^2} \sum_{n=1}^{\infty}\sum_{m=1}^{\infty} \sin\frac{n\pi}{\ell} x_1 \sin\frac{m\pi}{\ell} x_2 \left(\int_0^{\ell} \sin\frac{n\pi}{\ell} x \sin\frac{m\pi}{\ell} x dx\right)$$

$$\left(\int_0^t\int_0^t e^{\left(\frac{-n\pi}{\ell}\right)^2(t-\tau_1)} e^{\left(\frac{-m\pi}{\ell}\right)^2(t-\tau_2)} d\tau_1 d\tau_2\right)$$

The variance of the first order solution will be computed as

$$\text{Var}[u_1(t,x)] = E[u_1(t,x) - Eu_1(t,x)]^2$$

$$= E\left[\left(\sum_{n=1}^{\infty}(I_{n,1}(t) - EI_{n,1}(t))\sin\frac{n\pi}{\ell}x\right)^2\right]. \tag{15}$$

To give

$$\text{Var}[u_1(t,x)] = \frac{4\hbar^2\sigma^2}{\ell^2}\sum_{n=1}^{\infty}\sum_{m=1}^{\infty}\sin\frac{n\pi}{\ell}x\sin\frac{m\pi}{\ell}x\left(\int_0^\ell \sin\frac{n\pi}{\ell}x\sin\frac{m\pi}{\ell}x\,dx\right)$$

$$\left(\int_0^t\int_0^t e^{\left(\frac{-n\pi}{\ell}\right)^2(t-\tau_1)}e^{\left(\frac{-m\pi}{\ell}\right)^2(t-\tau_2)}d\tau_1 d\tau_2\right)$$

$$= \frac{2\left(1-e^{-\pi^2 t}\right)^2 \hbar^2 \sin[\pi x]^2}{\pi^4}.$$

In this manner, we can have more results of $E[u_m(t,x)]$ and $\text{Var}[u_m(t,x)]$ obtained at m=2,3,4,...

The final expression of mean of the 3rd order solution will be

$$E[u(t,x)] = E\left[\sum_{m=0}^{M} u_m(t,x)\right]$$

$$E[u(t,x)] = u_0(t,x) + E[u_1(t,x)] + E[u_2(t,x)] + E[u_3(t,x)]$$

$$E[u(t,x)] = \frac{1}{9}(9e^{t\beta n} + \frac{6e^{-\pi^2 t}\hbar\left(3\left(-1+e^{t(\pi^2+\beta n)}\right)\pi(\pi^2+2\beta n) - 8\varepsilon + 8e^{t(\pi^2+2\beta n)}\varepsilon\right)}{\pi^3 + 2\pi\beta n}$$

Solving Nonlinear Stochastic Diffusion Models With Nonlinear Losses

$$+\left(e^{-\pi^2 t}h^2\left(-128\beta n\varepsilon^2-\left(\pi^2+3\beta n\right)\left(9\pi^2\beta n\left(\pi^2+2\beta n\right)+72\pi\beta n\varepsilon\right.\right.\right.$$

$$\left.-16\varepsilon\left(3\pi^3+6\pi\beta n+8\varepsilon\right)\right)+e^{t\beta n}\left(128e^{t\left(\pi^2+2\beta n\right)}\beta n\varepsilon^2+\left(\pi^2+3\beta n\right)\right.$$

$$\left.\left.\left.\left(9e^{\pi^2 t}\pi^2\beta n\left(\pi^2+2\beta n\right)+72e^{t\left(\pi^2+\beta n\right)}\pi\beta n\varepsilon-16\varepsilon\left(3\pi^3+6\pi\beta n+8\varepsilon\right)\right)\right)\right)\right)$$

$$\Big/\left(\pi^2\beta n\left(\pi^2+2\beta n\right)\left(\pi^2+3\beta n\right)\right)\sin[\pi x].$$

Since $u(t,x) = \sum_{i=1}^{N} u_i(t,x)$

Then the final expression of the variance of the 2nd order solution will be

$$\text{Var}\left(\sum_{i=1}^{N}u_i(t,x)\right)=\sum_{i=1}^{N}\text{Var}\left[u_i(t,x)\right]+\left(\sum_{i=1}^{N}\sum_{j\neq i}^{N}\text{Cov}\left[u_i(t,x),u_j(t,x)\right]\right)$$

$$\text{Var}\left[u(t,x)\right]=\text{Var}\left[u_1(t,x)\right]+\text{Var}\left[u_2(t,x)\right]+2\text{Cov}\left[u_1(t,x),u_2(t,x)\right]$$

$$\text{Var}\left[u(t,x)\right]=h^2 t\left(4.+h\left(4.-1.7149t\right)+h^2\left(1.-0.735t+0.12t^2\right)\right)$$

$$\text{Var}\left[u(t,x)\right]=\frac{1}{\pi^8}4h^2\sin[\pi x]^2\left(3\left(-1+e^{-\pi^2 t}\right)^2\pi^4\right.$$

$$-1\Big/\left(\pi^2+\beta n\right)2e^{-2\pi^2 t}\pi^3\left(2\left(-1+e^{\pi^2 t}\right)^2\pi^2+2\left(-1+e^{\pi^2 t}\right)^2\beta n\right.$$

$$\left.+\left(-1+e^{\pi^2 t}\right)\left(-1+e^{t\left(\pi^2+\beta n\right)}\right)\pi\varepsilon-\left(-1+e^{t\left(\pi^2+\beta n\right)}\right)\pi^3 t\varepsilon\right)$$

$$-2e^{-2\pi^2 t}\left(\pi\left(-4\left(-1+e^{\pi^2 t}\right)^2 h^2\pi + \frac{1}{\beta n\left(\pi^2+\beta n\right)}\left(1-e^{\pi^2 t}+\pi^2 t\right)\right.\right.$$

$$\left.\left.\left(\left(-1+e^{t\beta n}\right)\pi^4 + e^{t\beta n}\left(-1+e^{\pi^2 t}\right)\beta n^2\right)\varepsilon\right) - \frac{16e^{2t\beta n}h^2\left(1-e^{\pi^2 t}+\pi^2 t\right)^2\varepsilon^2\sin[\pi x]^2}{\pi^2}\right)\right).$$

RESULT ANALYSIS

1D Diffusion Model Results

Figures 1 and 2 show the plots of the $\hbar$-curves for the fourth order variance and mean approximations respectively for different values of time t at a(t) = 1/2, $\lambda = 1$, $\varepsilon = 10^{-2}$ and $x_0=0.5$ on the time interval [0,]">2]. According to these -curves, it is easy to discover that the valid region of $\hbar$ is a horizontal line segments, thus $\hbar = -092$ Figures 3 and 4 show the comparison of the expectation and variance as a function of time using HAM and WHEP which uses the Wiener Hermite expansion and perturbation technique to solve a class of nonlinear partial differential Equations with a perturbed nonlinearity "techniques and good agreement is obtained.

The mean and variance results of the WHEP technique are obtained from [5] as:

$$E[x(t)] = x_0 e^{at} - \varepsilon\frac{\left(e^{at}-1\right)\left(e^{at}\left(2a(x_0)^2+\lambda^2\right)-\lambda^2\right)}{2a^2}$$

$$\mathrm{Var}[x(t)] = \frac{\lambda^2}{2a}\left(e^{2at}-1\right) - 2\varepsilon\frac{\lambda^2 x_0}{a^2}\left(e^{at}-1\right)^2\left(e^{at}+1\right) + 2\varepsilon^2\frac{\lambda^2(x_0)^2}{a^3}\left(e^{at}-1\right)^3\left(e^{at}+1\right)$$

The effect of ε on the variance is shown in Figure 5. The variance is plotted with time for different values of ε. The peak variance decreases in magnitude with the increase of ε. Also, the time of the peak variance decreases with the increase of ε.

2D Diffusion Model Results

In the following figures, results of the solution of 2D stochastic quadratic nonlinear diffusion model using HAM technique are shown at $\sigma = 1, \ell = 1, \beta_n = -1, n = 1, \varepsilon = 1, \phi(x) = \sin\frac{n\pi}{\ell}x$.

Figure 6 shows the Plot of $\hbar$-curve of third order approximation of mean for different values of time t and space variable x at $\sigma = 1, \ell = 1, \beta_n = -1, \varepsilon = 1, n = 1, \phi(x) = \sin\frac{n\pi}{\ell}x$,.Figure 7 shows the plot of $\hbar$-curve of

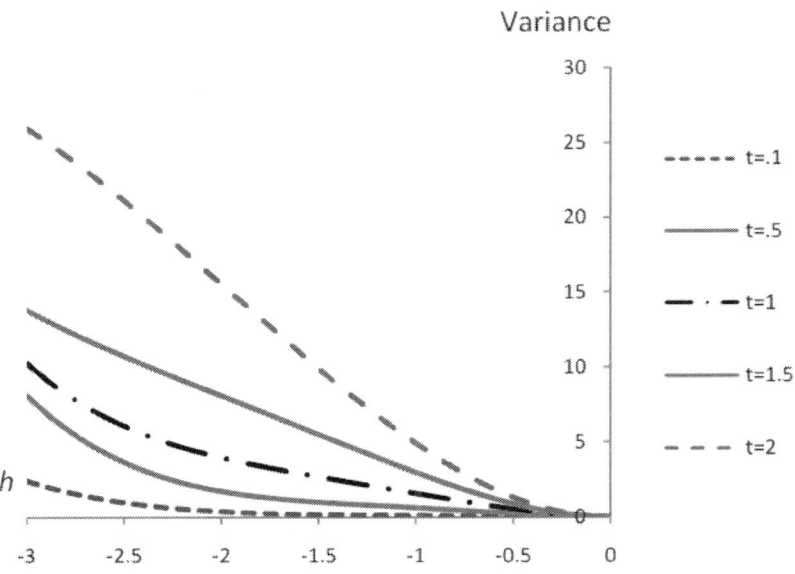

Figure 1: The change of variance of the solution x(t) with parameter $\hbar$ at different t values.

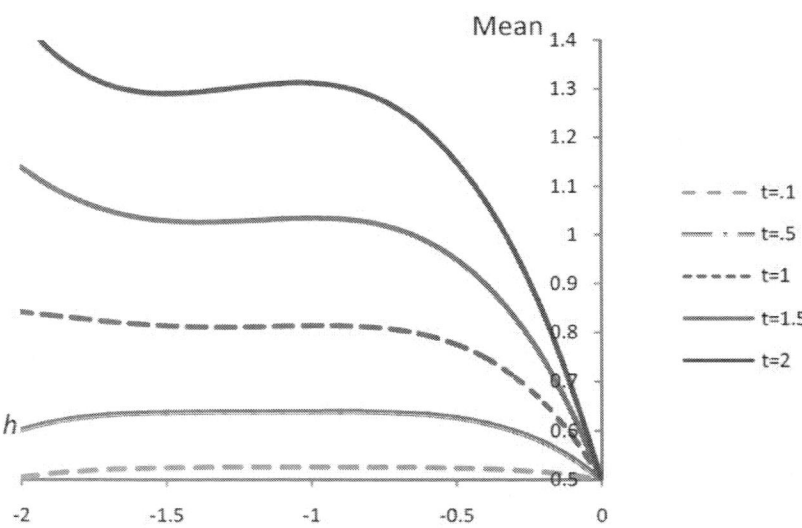

Figure 2: The change of mean of the solution x(t) with parameter $\hbar$ at different t values.

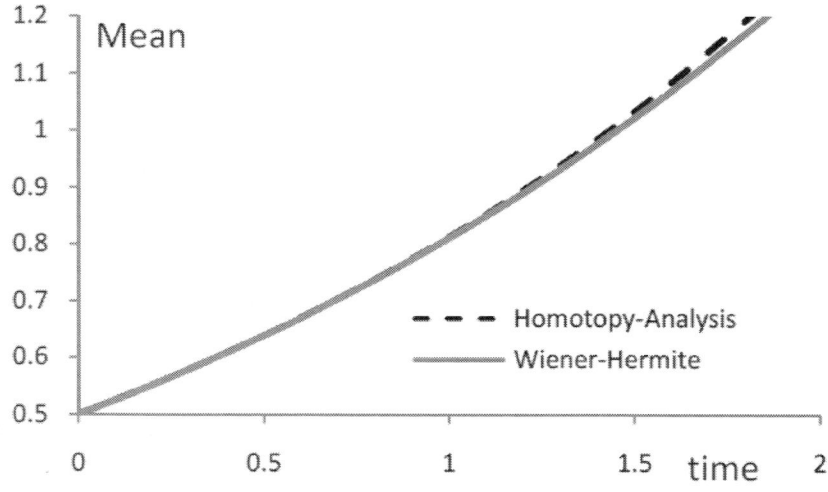

Figure 3: Comparison of the expectation obtained by using HAM at $\hbar = -0.9$ for the 1D problem and WHEP [18].

Solving Nonlinear Stochastic Diffusion Models With Nonlinear Losses

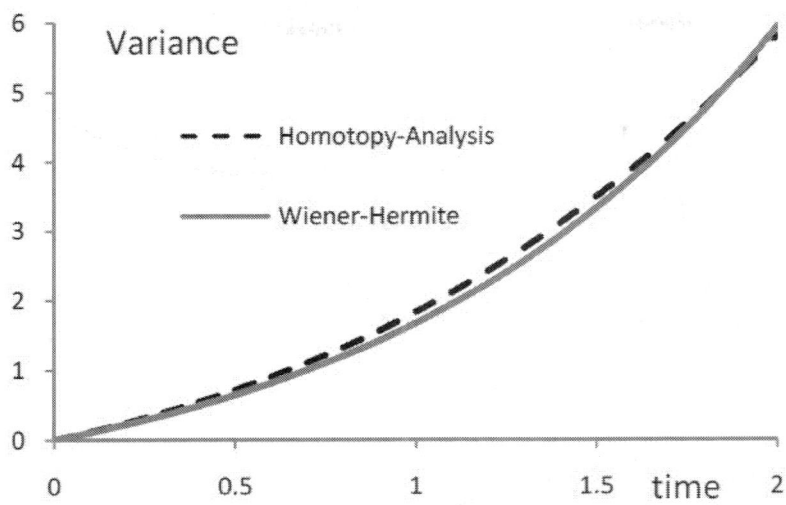

Figure 4: Comparison of and the variance obtained by using HAM method at $\hbar$ =-0.92 for the 1D and WHEP [18].

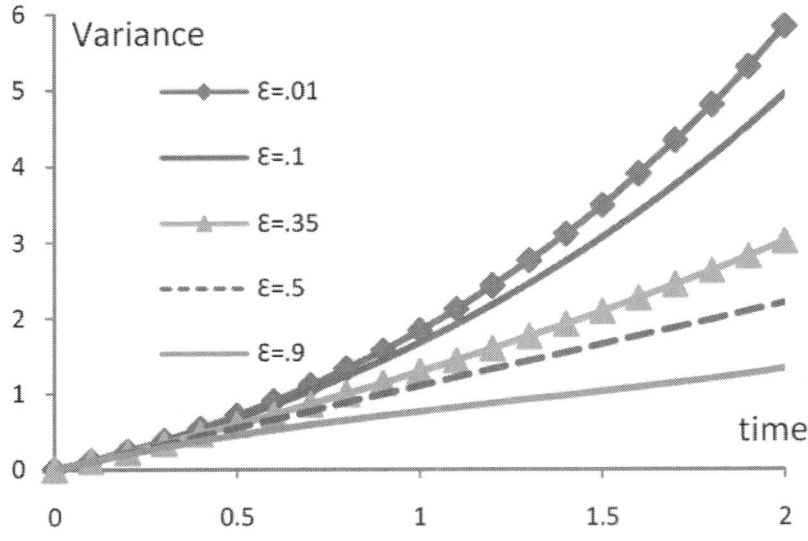

Figure 5: The effect of ε on Var[x(t)].

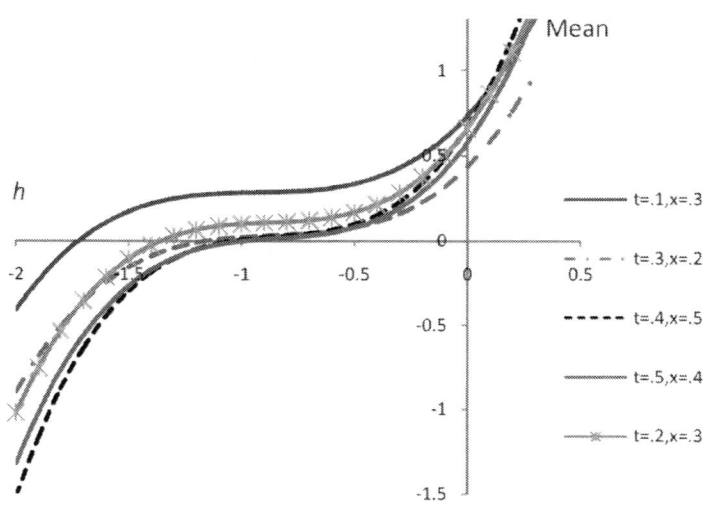

Figure 6: The change of the mean u with parameter $\hbar$ at different t, x values.

third order approximation of mean for different β_n values. According to these $\hbar$-curves, it is easy to discover that the valid region of $\hbar$ is a horizontal line segments, thus $\hbar = -096$. Figures 8 and 9 show the plot of mean and variance with time for different ε values.

Figure 10 shows the comparison between the mean of the first, the second and the third order approximations. Figure 11 shows the comparison between the variance of the first and second order approximations.

CONCLUSION

This paper shows that the HAM technique constitutes a powerful tool for constructing approximate solutions for the stochastic process for random diffusion models with nonlinear perturbations where uncertainty is considered by means of an additive term defined by white noise. The HAM method is employed to give a statistical analytic solution for stochastic 1D and 2D diffusion models. Different from all other analytic methods, the HAM provides us with a simple way to

Solving Nonlinear Stochastic Diffusion Models With Nonlinear Losses

adjust and control the convergence region of the series solution by means of the auxiliary parameter ℏ. Thus the auxiliary parameter ℏ plays an important role within the frame of the HAM which can be determined by the so called ℏ-curves. The solution obtained by means of the HAM is an infinite power series for appropriate initial approximation, which can be, in turn, expressed in a closed form. The accuracy for the method is verified on 1D diffusion model by comparisons with WHEP technique and good agreements are obtained. As shown in Figures 1 and 2, we can see that the valid ℏ region in the 1D example is $-0.9 < \hbar < -1.4$ and in 2D example the interval is $-0.9 < \hbar < -1.1$, as shown in Figure 6. The results demonstrate reliability and efficiency of the HAM method. Since HAM was used to solve only deterministic problems, we

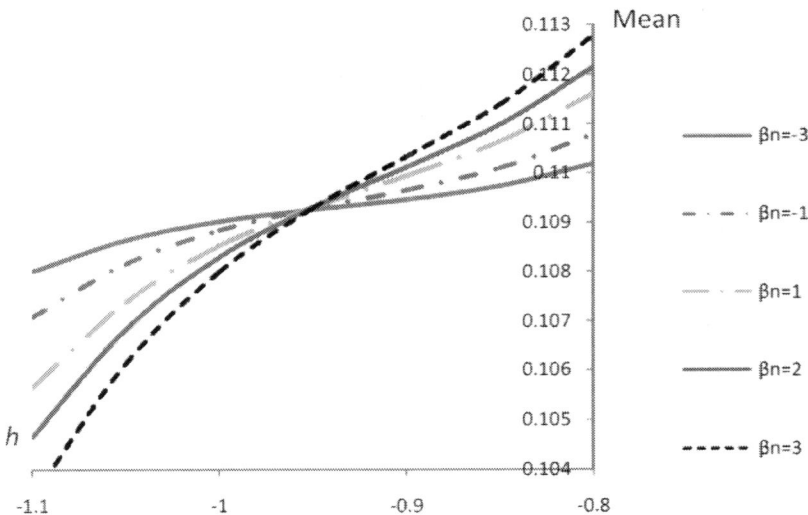

Figure 7: The change of the mean u with parameter ℏ at different β_n values, $\varepsilon = 1$, t = x = 0.1.

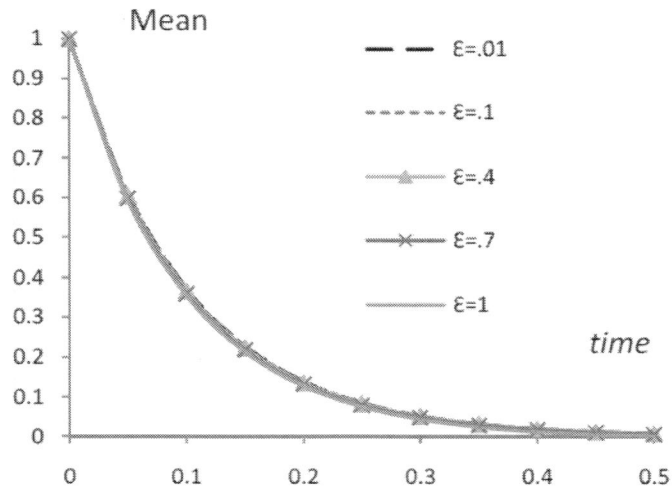

Figure 8: The change of the mean u with time t at different ε values, x = 0.1, 1 β_n =-1, $\hbar$ =-0.96.

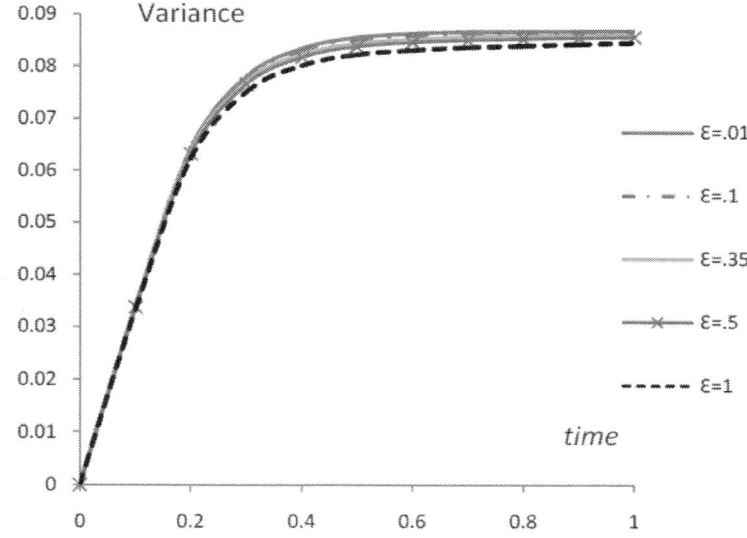

Figure 9: The change of the variance u with time t at different ε values, x = 0.1, β_n =-1, $\hbar$ =-0.96.

Solving Nonlinear Stochastic Diffusion Models With Nonlinear Losses

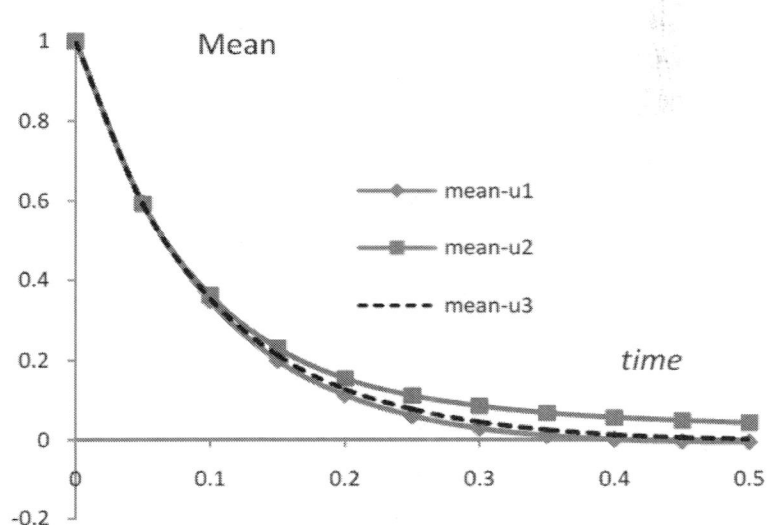

Figure 10: Mean comparison between first u_1, second u_2 and third order u_3 approximations with time t at, x = 0.1, β_n=-1 $\hbar$ =-0.96.

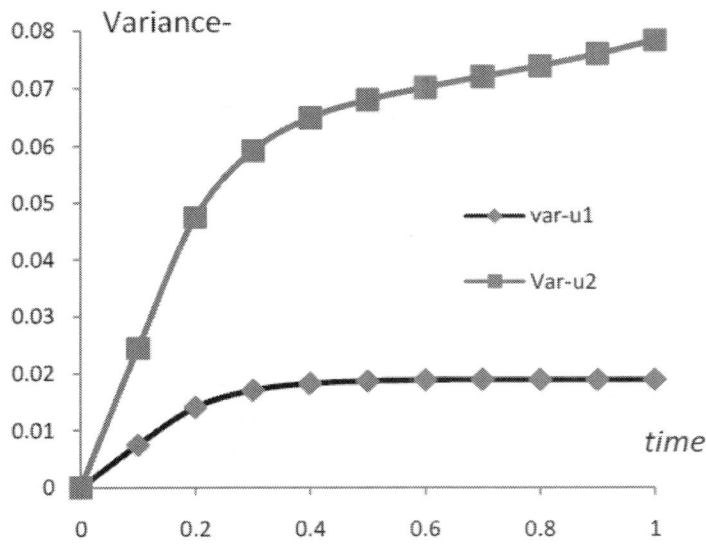

Figure 11: Variance comparison between first and second approximations u_1, u_2 with time t at, x = 0.1, β_n=-1, $\hbar$ =-0.96.

can say that this is the first time to apply HAM method on stochastic problems and we found that it's easier than WHEP and more general than HPM since HPM is a special case of HAM obtained at $\hbar = -1$ and its results are accurate.

REFERENCES

1. M. A. El-Tawil and A. S. Al-Jihany, "On the Solution of Stochastic Oscillatory Quadratic Nonlinear Equations Using Different Techniques, a Comparison Study," Topological Methods in Nonlinear Analysis, Vol. 31, No. 2, 2008, pp. 315-330.
2. M. A. El-Tawil and N. A. Al-Mulla, "Using Homotopy WHEP Technique for Solving A Stochastic Nonlinear Diffusion Equation," Mathematical and Computer Modelling, Vol. 51, No. 9, 2010, pp. 1277-1284.
3. J. C. Cortes, J. V. Romero, M. D. Rosello and C. Santamaria, "Solving Random Diffusion Models with Nonlinear Perturbations by the Wiener-Hermite Expansion 617 Method," Computers & Mathematics with Applications, Vol. 61, No. 8, 2011, pp. 1946-1950.
4. C. Q. Dai and J. F. Zhang, "Application of He's Exp-Function Method to the Stochastic mKdV Equation," International Journal of Nonlinear Sciences and Numerical Simulation, Vol. 10, No. 5, 2009, pp. 675-680.
5. M. El-Beltagy and M. El-Tawil, "Toward a Solution of a Class of Non-Linear Stochastic Perturbed PDEs Using Automated WHEP Algorithm," Applied Mathematical Modeling, Vol. 37, No. 12-13, 2013, pp. 7174-7192. http://dx.doi.org/10.1016/j.apm.2013.01.038
6. S. J. Liao, "The Proposed Homotopy Analysis Technique for the Solution of Nonlinear Problems," Ph.D. Thesis, Shanghai Jiao Tong University, Shanghai, 1992.
7. S. J. Liao, "Notes on the Homotopy Analysis Method: Some Definitions and Theories," Communications in Nonlinear Science Numerical Simulation, Vol. 14, No. 4, 2009, pp. 983-997.
8. G. Adomian, "A Review of the Decomposition Method and Some Recent Results for Nonlinear Equations," Computers and Mathematics with Applications Vol. 21, No. 5, 1991, pp. 101-127.
9. J. H. He, "Homotopy Perturbation Method: A New Nonlinear Analytical Technique," Applied Mathematics and Computation Vol. 135, No. 1, 2003, pp. 73-79.
10. T. Hayat and M. Sajid, "Analytic Solution for Axisymmetric Flow and Heat Transfer of a Second Grade Fluid Past a Stretching Sheet," International Journal of Heat and Mass Transfer Vol. 50, No. 1-2, 2007, pp. 75-84.

11. S. Abbasbandy, "Soliton Solutions for the 5th-Order KdV Equation with the Homotopy Analysis Method," Nonlinear Dynamics, Vol. 51, No. 1-2, 2008, pp. 83-87.
12. Y. P. Liu and Z. B. Li, "The Homotopy Analysis Method for Approximating the Solution of the Modified Korteweg-de Vries Equation" Chaos Solitons and Fractals, Vol. 39, No. 1, 2009, pp. 1-8.
13. Y. Bouremel, "Explicit Series Solution for the Glauert-Jet Problem by Means of the Homotopy Analysis Method," Communication in Nonlinear Science Numerical Simulation, Vol. 12, No. 5, 2007, pp. 714-724.
14. A. Molabahrami and F. Khani, "The Homotopy Analysis Method to Solve the Burgers-Huxley Equation," Nonlinear Analysis Real World Applications Vol. 10, No. 2, 2009, pp. 589-600.
15. S. Abbasbandy, E. Magyari and E. Shivanian, "The Homotopy Analysis Method for Multiple Solutions of Nonlinear Boundary Value Problems," Communications in Nonlinear Science and Numerical Simulation Vol. 14, No. 9-10, 2009, pp. 3530-3536.
16. H. N. Hassan and M. A. El-Tawil, "Solving Cubic and Coupled Nonlinear Schrödinger Equations Using the Homotopy Analysis Method," International Journal of Applied Mathematics and Mechanics, Vol. 7, No. 8, 2011, pp. 41-64.
17. H. N. Hassan and M. A. El-Tawil, "An Efficient Analytic Approach for Solving Two-Point Nonlinear Boundary Value Problems by Homotopy Analysis Method," Mathematical Methods in the Applied Sciences, Vol. 34, No. 8, 2011, pp. 977-989.

CITATION

Aisha A. Fareed, Hanafy H. El-Zoheiry, Magdy A. El-Tawil, Mohammed A. El-Beltagy, Hany N. Hassan Solving Nonlinear Stochastic Diffusion Models with Nonlinear Losses Using the Homotopy Analysis Method Applied Mathematics Vol.5 No.1(2014), Article ID:41820,13 pages DOI:10.4236/am.2014.51014

Stochastic Process Optimization Technique

Hiroaki Yoshida[1], Katsuhito Yamaguchi[2], and Yoshio Ishikawa[1]
[1]College of Science and Technology,
Nihon University, Chiba, Japan
[2]Junior College, Nihon University,
Chiba, Japan

ABSTRACT

The conventional optimization methods were generally based on a deterministic approach, since their purpose is to find out an accurate solution. However, when the solution space is extremely narrowed as a result of setting many inequality constraints, an ingenious scheme based on experience may be needed. Similarly, parameters must be adjusted with solution search algorithms when nonlinearity of the problem is strong, because the risk of falling into local solution is high. Thus, we here propose a new method in which the optimization problem is replaced with stochastic process based on path integral techniques used in quantum mechanics and an approximate value of optimal solution is calculated as an expected value instead of accurate value. It was checked through some optimization problems that this method using stochastic process is effective. We call this new optimization method "stochastic process optimization technique (SPOT)". It is expected that this method will enable efficient optimization by avoiding the above difficulties. In this report, a new optimization method based on a stochastic process is formulated, and several calculation examples are shown to prove its effectiveness as a method to obtain approximate solution for optimization problems.

INTRODUCTION

Optimization methods are useful to system design, and especially the applications to engineering problems are expected. In order to obtain the accurate optimal solution, generally the conventional optimization methods were based on a deterministic approach. However, when design variables have many inequality restriction conditions, in order to search for a solution efficiently, it is necessary to adjust a parameter with trial and error. Moreover, if the problem is complex, the risk of falling into a local solution is also high.

Our method obtains an approximate value of an optimal solution using stochastic process. The optimization methods using stochastic process like simulated annealing (SA) [1] are found. However, in these methods, stochastic process is used as a means of searching for an accurate solution efficiently. Our method is based on probability theory, and essentially differs from the concepts of the conventional optimization methods. The purpose of the conventional optimization is to obtain the accurate optimal solution, but that of our method is to obtain the approximate value of the optimal solution.

We showed that the approximate value of the optimal solution could be obtained using stochastic process [2] [3] . And even if it was an approximate value of the optimal solution obtained by this method, it was shown that it is useful in engineering [4] . Furthermore, it was shown that this method can apply to general engineering optimization problems containing static design variables and dynamic design variables [5] .

As mentioned above, it was shown that the approximate value of the optimal solution can be obtained by this method using stochastic process. The aim of this paper is not to show the results of the applications of our method, but to bring our original idea to a wide reading audience.

Thus, we propose a new method in which an optimization problem is replaced with a stochastic process based on path integral techniques used in quantum mechanics and an approximate value of an optimal solution is calculated as an expected value instead of an accurate value. We call this new optimization method the "stochastic process optimization technique (SPOT)". It is expected that this method will enable efficient optimization by avoiding the above difficulties.

Since expected values are stochastic average values, a stochastic approach is not appropriate to obtain an accurate solution as a deterministic approach. However, its characteristics are simple algorithms and the fewer number of manual parameters. The resultant solution is approximate enough to a global solution of engineering problems whose solution spaces are expected to be relatively smooth. The approximate solution can be an initial value to obtain an accurate solution with a deterministic process.

In this paper, a new optimization method is formulated based on a stochastic process. Some calculation examples are introduced to show how effective the approximate solutions are for optimization problems. The examples are a simple benchmark test, a classic line of swiftest descent problem, and an optimized glide problem of a hang glider as an introductory engineering optimization problem.

PATH INTEGRAL AND OPTIMIZATION

According to the principle of least action, Newtonian motion of a mass point makes the action integral minimum:

$$I = \int_{t_1}^{t_2} L(x_i, dx_i/dt, t) \, dt ,\tag{1}$$

where $L(x_i, dx_i/dt, t)$ is the Lagrange function. Also, the Euler-Lagrange equation obtained with variation of I is equal to an equation of Newtonian motion. On the other hand, the motion of a particle is determined stochastically according to the quantum mechanics. A motion on the classical mechanics, which keeps I minimum, occurs at the "highest possibility". However, other motions can occur at certain probabilities.

This results in some particular phenomena, such as the tunnel effect, which never occur on the classical mechanics.

The well-known simulated annealing (SA) method adopts such a quantum mechanical approach to the solution of optimization problems. SA introduces a probability distribution that reaches its maximum at an optimal value. This method numerically searches the optimal value (peak of the probability distribution) with a temperature parameter using the probability distribution that reaches its maximum of an optimal value instead of directly obtaining a variational problem of a performance index I. This method searches solution spaces in the large, which include solutions that never obtained with the variational method. Therefore, one advantage is a low probability of reaching local solutions. On the other hand, there are some disadvantages. For example, the operation of a cooling schedule, which is a temperature parameter, is difficult. In addition, it may take much time in calculation to converge a solution.

R. Feynman constructed the "path integral" theory in 1948 and showed that it is equivalent to the quantum mechanics. The theory provides a countless number of paths that meet given boundary conditions, and the prob- ability (probability amplitude) of each particle taking its path is expressed in the form of multiple integral [6] . A stochastic behavior is the nature of quantum mechanics. It is therefore impossible to predict a measured physical quantity deterministically under a certain condition. Expected values only can be predicted. Thus, the expected value of a physical quantity is expressed in the form of multiple integral in the path integral method. Furthermore, the multiple integral calculation of an expected value can fit into numerical calculation with the Monte Carlo method.

In this paper, we propose a new numerical calculation method for optimization problems based on the quantum mechanics rather than SA. It means that we show a method of obtaining an approximate solution of an optimal solution as an expected value. An expected value is predicted to be located close to an optimal solution in a continuous probability distribution. This method can thus be applied to various engineering optimal problems.

Stochastic Process Optimization Technique

The introduced probability distribution P here is basically the same as that in SA, and to obtain the minimum of performance function I, the following equation is set:

$$P = e^{-I/h}. \qquad (2)$$

In SA, the peak of P is searched with $h \to 0$. On the other hand, we automatically generate a state variable of various time histories at probability P as finite value h. According to the generated frequency, the stochastic average (expected value) of the quantity to be obtained is calculated.

FORMULATION OF SPOT

According to the path integral method, the path of motion for a naturally stochastic particle such as photon and electron is obtained from an expected value as a path that keeps the action integral I minimum. We replace the action integral I with a performance index to recognize the motion of a general object as a stochastic process. Thus, we formulate the path integral method for a naturally stochastic particle in physics as an optimal method of obtaining a function that keeps the performance function I minimum. Note that since this performance function is stochastic, an optimal solution is obtained as an expected value, which results in an approximate solution instead of an accurate solution.

According to the path integral method, the action integral [x(t)] described in the above section is replaced with a performance function to define probability distribution P as shown in Equation (3). Here, x(t) does not always show a path, but shows a general variable to be optimized. Also, A is the normalization constant.

$$P = \frac{1}{A}\exp(-I/h), \qquad (3)$$

where h is the Plank constant in quantum mechanics. A stochastic mechanics system matches a classical mechanics system as h → 0. However, in this situation, h is an arbitrary parameter that gives a fluctuation of solution. It means that h is a parameter to define a distribution of P. The larger h is, the wider the distribution is. Also, A is the normalization constant to make the summation of the probability 1. To define the probability distribution as shown in Equation (3), the expected value is obtained from the following:

$$\langle x \rangle = \int_{-\infty}^{\infty} \left[x(t) \cdot P \right] Dx(t), \qquad (4)$$

where $Dx(t) = dx_0 dx_1 \cdots dx_N$ shows the summation for the paths.

Here, the summation for the paths is the integral for all the countless paths (multiple integral), which for example connect each point x_i with lines at each time point t_i ($0 \leq i \leq N$) divided by N in one-dimensional motion. Figure 1 shows some paths that are generated in a probability distribution of Equation (3) when a function defined in Equation (1) is a performance index in one-dimensional upcast motion.

The horizontal axis and vertical axis show time and vertical positions, respectively. The better performance values the paths have, in other words, the closer to the accurate solution (red line) the values are, the higher the density of the paths is. Also, there are some paths that never exist with a deterministic method. The optimal solution is the expected value of all the paths. In this case, that solution is an approximate solution of an accurate solution. The expected solution that is thus obtained shows a path of the highest probability for move of a stochastic behavioral particle, such as an electron, between two points. When those paths are replaced with arbitrary functions, approximate solutions that make the performance function values minimum are obtained.

This idea can be extended to obtain an expected value of a m-dimensional functional that consists of m×N functions of time in the following way. First, independent variable t is divided by N, and $x_k^{(i)}$ values

Stochastic Process Optimization Technique

at time $t_i (0 \leq i \leq N)$ are randomly chosen. Next, a path that connects these points is set to the time history of variable x_k (Figure 2).

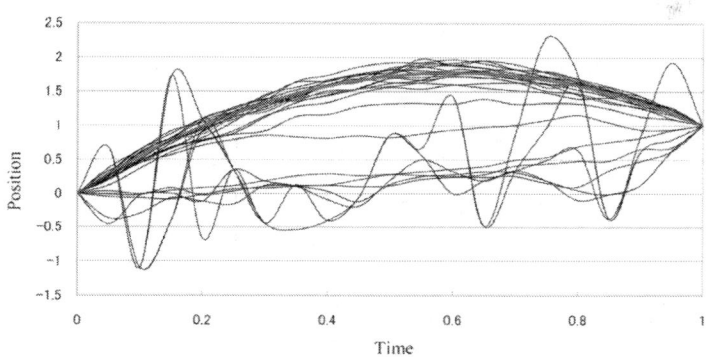

Figure 1: Paths generated by a probability distribution.

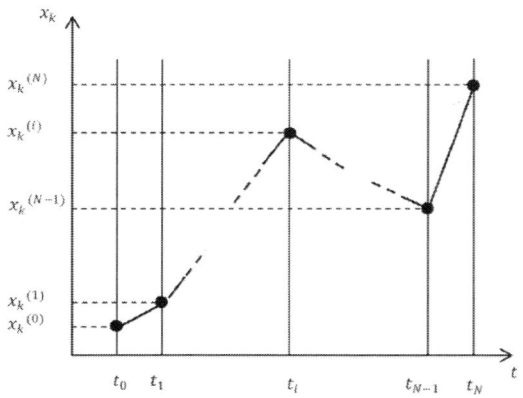

Figure 2: Time history of a variable.

Note that the subscript k shows the k-th variable $(0 \leq k \leq m)$. From Equation (4), the expected value $\langle x_k^{(i)} \rangle$ of x_k at t_i $(0 \leq i \leq N)$ is expressed as follows:

$$x_k^{(i)} = \int_{-\infty}^{\infty} \left[x_k^{(i)} \cdot P \right] Dx^{(0)}(t) Dx^{(1)}(t) \cdots Dx^{(N)}(t),$$
$$Dx^{(i)} = dx_1^{(i)} \cdot dx_2^{(i)} \cdots dx_m^{(i)}. \tag{5}$$

The integral range is from $-\infty$ to ∞ in Equation (5) according to the form of the path integral. The range in actual calculation is between the upper and lower limits of each variable. Also, Equation (5) is formulated to set the values according to the probability distribution for variable $x_k^{(i)}$ at time 0 and N. However, a variable may be fixed in some problems. The end points should be set in accordance with the boundary conditions.

Note that since Equation (5) shows the multiple integral, the Monte Carlo method can be applied to actual numerical calculations.

CALCULATION ALGORITHM FOR SPOT

First, a performance index is determined, and if necessary, variables are discretized for numerical calculation. Next, the values of all variables are randomly chosen to generate solutions. Generation probabilities of variables should agree with the probability distribution of Equation (3). Thus, a solution with a fine performance value is generated at a high frequency, and a solution with a poor performance index value is generated at a low frequency. Also, the algorithm is based on the Markov process so that future solutions will have no dependency on the former solutions. Lastly, expected values are calculated using all the solutions generated with a probability distribution to obtain an approximate solution.

The procedure of calculation is shown in Figure 3.

1. The values of all the variables are generated randomly to obtain initial solutions.
2. Time point t_i is chosen according to a random number, and function value x_i at t_i is randomly rewritten to generate a different solution.

Stochastic Process Optimization Technique

3. The performance index value of the newly generated solution is calculated.
4. The adoption or rejection of this solution is decided to generate solutions depending on the probability distribution of Equation (3).

The procedure from Steps 2) to 4) is a pure explanation of the Metropolis method [7] to merely conduct the multiple integral. This method has an advantage of no necessity to obtain the normalization constant: A in Equation (3). A prime characteristic of this method is to calculate the expected values in Step 5).

5. The expected value is calculated.
6. If the final condition is met, the calculation ends. If not, it returns to the generation of solutions.

The final condition can be a certain range of variation of an expected value. However, in this paper, the final condition is the number of iterations, K_{max}. This is because the generation of solutions during the calculation is a Markov process, and therefore, the still more iterations of calculation can result in fine performance index values if the convergence of expected values is not enough.

Equation (3) for probability distribution in this paper is the same as that in SA. SA is a deterministic approach to search only optimal solution at $h \to 0$. On the other hand, is finite and an optimal solution is obtained as an expected value (stochastic average value) in this paper. It means that the two approaches are based on fundamentally different concepts. In the concrete, SA starts a search with an initial solution for candidate solutions according to the probability distribution. It adopts a solution of the best performance function value as an optimal solution. On the other hand, our method determines an approximate solution to an optimal solution as an expected value with all the solutions generated according to the probability distribution. In this method, there is no process of the so-called cooling schedule which is in SA.

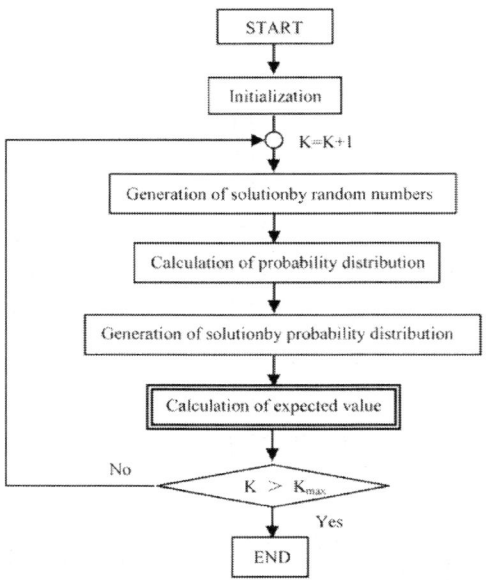

Figure 3: Flow chart of calculation.

CALCULATION EXAMPLE

This section describes examples of numerical calculation with SPOT to demonstrate the validity. First, SPOT is applied to a problem where the minimum value of a multi-peaked function is obtained without staying at other minimal values and the minimum value is always only one with no dependency on the initial value. Next, we apply SPOT to a problem of obtaining the minimum value of a functional to show the validity in a physical optimization problem. Lastly, we apply SPOT to an engineering optimal problem on the glide of a hang glider to show the high applicability of an approximate solution by SPOT.

We already applied SPOT to some engineering problems [8] [9].

Stochastic Process Optimization Technique

Minimization Problem of Bivariate Function

First, we apply SPOT to a problem of obtaining the minimum value of a bivariate function expressed in Equation (6). Here, performance function I is the function value f(x,y) as follows:

$$f(x,y) = \{\cos(2\pi x) + \cos(2.5\pi x) - 2.1\} \times \{2.1 - \cos(3\pi y) - \cos(3.5\pi y)\}, \quad 0.0 \le x \le 3.0, \ 0.0 \le y \le 1.5 \quad (6)$$

Figure 4(a) shows the solution space of Equation (6). The set of f(x,y) that minimizes this function is (0.439, 0.306). This function has multiple minimal values and thus is a multi-peaked function. The contours are shown in the bottom plane, and Figure 4(b) shows the top view.

This example is a problem for obtaining the minimum value of a bivariate function. Therefore, the variables are not discretized, but the values of two variables are generated according to the probability distribution (3) defined by the performance values to obtain an expected value. The closed dots in Figure 4(a) show multiple initial values chosen to check the dependency on the initial values. Also, the cross marks in Figure 4(b) show the minimum values obtained with this method.

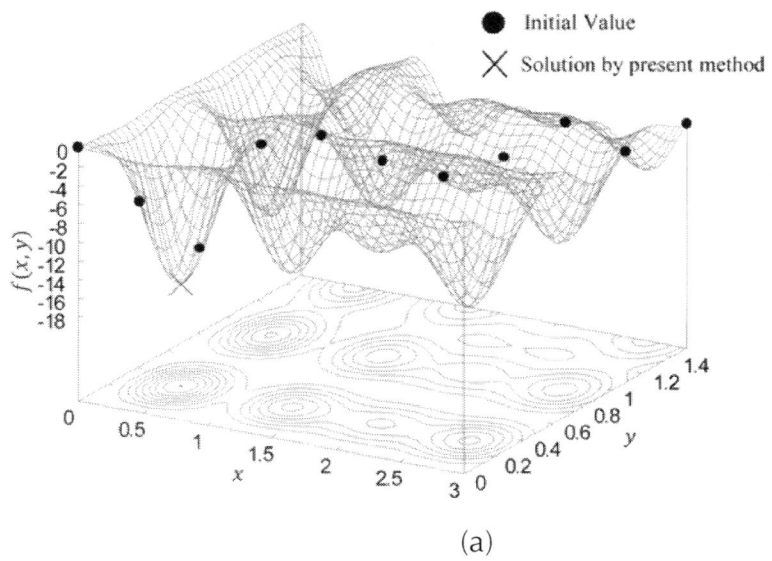

(a)

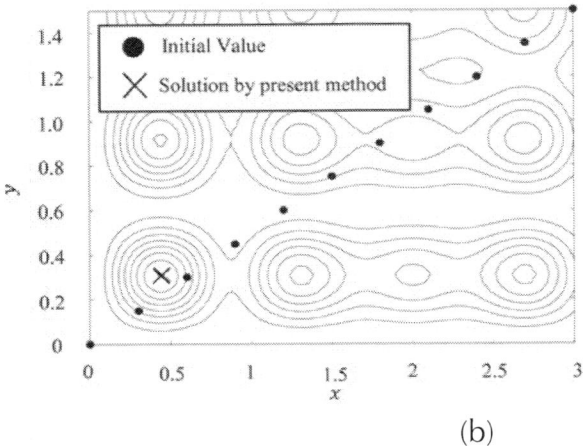

Figure 4: (a) Function space; (b) Contour lines of solution space.

If calculation starts with any initial value marked in the closed dot, the resultant values do not stay at any of the many minimal values in the solution space, but all the values concentrate near the minimum value. Since every solution generated in each calculation is obtained stochastically with this method, the calculation results are not influenced by the initial value in principle. However in reality, the number of calculations is finite. Therefore, the expected value as the minimum value may have a variation. The minimum values in Figure 4(b) have a little variation though it may be invisible. This variation is caused by the limited number of calculations.

Note that it took about 18 seconds to perform one million times of calculations on a 1.4 GHz Pentium 4 machine.

Line of Swiftest Descent Problem

Next, we applied SPOT to a problem on brachistochrone as one of the classical optimization problems to obtain a function. This problem is demonstrated to obtain a mass point path that shows the brachistochrone between a fixed start point and an end point at different altitudes under the uniform gravitation field. Here, performance function I

Stochastic Process Optimization Technique

is time T until the mass point reaches the end point as expressed in the following equation:

$$t = \frac{1}{\sqrt{2g}} \int_0^{y_f} \frac{\sqrt{1+(dx/dy)^2}}{\sqrt{y}} dy.$$

(7)

The boundary conditions are as follows:

Initial condition: $y(0) = x(0) = 0$, final condition: $y(t_f) = 1.0 x(t_f) = \pi/2$.

In this problem, we discretized the vertical axis y at an even division, and generate the values of the horizontal axis x corresponding to y according to the probability distribution of Equation (3) defined by the performance index of Equation (7) to obtain an expected value.

Figure 5 shows the paths obtained with SPOT in the closed dots and the analytic solution in the solid line. It took about 438 seconds on a Pentium 4 machine to perform 20 million times of iterations.

SPOT is to obtain an approximate solution of an optimal solution. However, the approximate solutions agree well with the analytic solutions. Thus, this method can be applied to general optimization problems where performance functions are expressed in functional.

Optimal Glide Problem of Hang Glider

Lastly, we apply SPOT to a problem of a hang glider that starts the fall and glide at an altitude of 12 m toward the ground to obtain its flight operation (time history of angle of attack) for the maximum down range (as shown in Figure 6).

This problem was firstly solved by Suzuki et al. [10] and their results were used to check the validity of the results of SPOT.

This problem is based on the former problem of [10] and the specifications of the hang glider used here are based on those of the optimal glider obtained in [10]. The shape of the hang glider is shown in Figure 7.

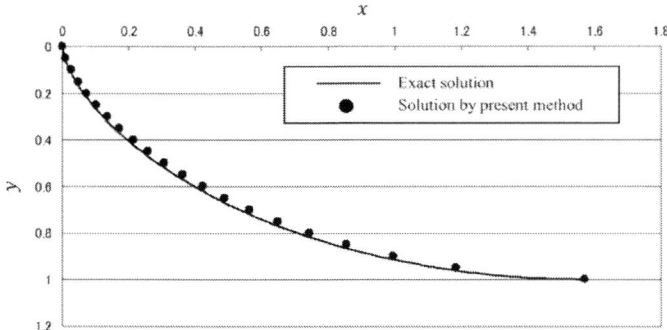

Figure 5: Line of swiftest descent.

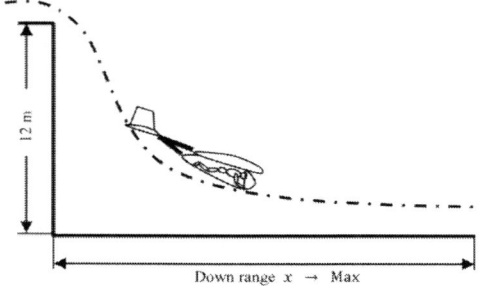

Figure 6: Flight trajectory optimization.

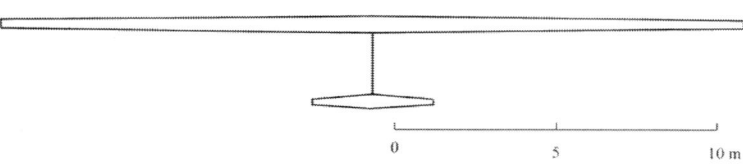

Figure 7: Top view of the hang glider.

Note that the performance index is the reciprocal number of the flight distance to set this problem to a minimal value problem. Also, the maximum value of the load factor is set to the restriction condition. If this condition exceeds the limitation during the flight, a penalty is added to the performance index value.

Stochastic Process Optimization Technique

The equation of motion is as follows:

$$\dot{V} = -(1/2m)\rho V^2 S_1 C_D^* - g\sin\gamma, \tag{8}$$

$$\dot{\gamma} = (1/2m)\rho V S_1 C_L^* - (g/V)\cos\gamma, \tag{9}$$

$$\dot{x} = V\cos\gamma \tag{10}$$

$$\dot{z} = V\sin\gamma \tag{11}$$

The main conditions are shown as follows:

Performance index:

$$I = 1/x_f + \sum_{n_i > n_{max}} n_i/x_f \tag{12}$$

Initial condition: $z_0 = 12.0\,\text{m}, V_0 = 5.0\,\text{m/s}, \gamma_0 = -3.5\,\text{deg}$.

Final condition: $z_f = 2.0\,\text{m}$.

Restriction condition: $n_{max} = 3.0$.

Specifications of hang glider: $S_1 = 14.6\,\text{m}^2$, $A_1 = 36.0$, $b_1 = 22.9\,\text{m}$, $\lambda_1 = 0.5$, $S_2 = 2.3\,\text{m}^2$, $A_2 = 6.0$, $\lambda_2 0.5$, $m_0 = 35.0\,\text{kg}$, $m_p = 60.0\,\text{kg}$.

The performance index value is the reciprocal number of the down range at the point where the hung glider reaches at an altitude of 2 m. First, the control variable, which is the angle of attack in this case, is discretized on the axis of time. Next, values of angle of attack at each time are specified with random numbers. The equations of motion are numerically calculated according to the time history of angle of attack to obtain the flight path. Then, the performance index value is determined as a reciprocal number of the down range. Therefore, the performance index values are obtained when the series of calculations ends with the final conditions being met.

Figure 8(a) and Figure 8(b) show the relation between the down range and the altitude, and that between the down range and the angle of attack, respectively.

This result may be an approximate solution close to the former optimization [9]. Note that it took about 15 hours on a Pentium 4 (1.4 GHz) machine to obtain the calculation result in this example that set the number of iterations to 3 millions.

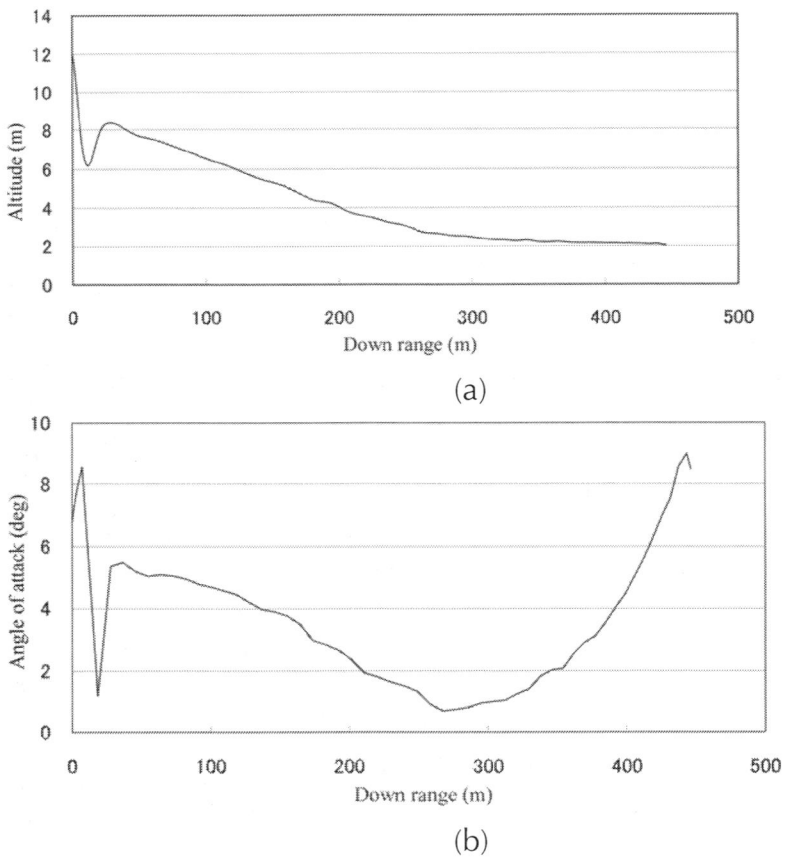

Figure 8: (a) Altitude vs. down range; (b) Angle of attack vs. down range.

DISCUSSION

We propose a new optimization method with a stochastic process called SPOT, and then demonstrate numerical calculations.

Generally, when proposing a new method, comparison of performance between the conventional method and a new method should be performed. However, since SPOT is the only method of obtaining the approximate value of the optimal solution, there is no sense in comparing them. Also fair comparison of computation time could not be performed.

As a result, SPOT has the following characteristics:

Advantages of Stochastic Process

SPOT results in an expected value from all the solutions generated by calculations. One of the advantages is to avoid influences from the initial value (a start point of the calculation) to the final expected value. The reason is that the generation of solutions during the calculation is a simple stochastic process, and the initial value is only the first-generated solution. As shown in Section 6.1, the calculation results all do not stay at local minimal values even though the calculation starts with any initial value, but the resultant values concentrate near the minimum value. Another advantage is to update an expected value from optimal solutions separately obtained under the same condition. This enables a parallel calculation and the recycling of the former calculation results. This characteristic is unique to SPOT and advantageous to other numerical calculation methods. For example, the calculation of a bivariate function, as shown in Section 6.1, gives the results starting with the multiple initial points as different minimum values. (The cross marks actually show almost the same point.) On the other hand, it is possible to obtain an expected value with all these expected values.

Also, SPOT seems to use asymptotic calculation in the algorithm. However, the former solutions are not used to generate the next solution. This means that the series of calculation is independent and thus the calculation time can easily be estimated with the number of calculations. Therefore, it is easy to estimate the calculation cost.

Independency on Experience

Another characteristic is that SPOT has only fluctuation parameter, h, set by operators. There may be a case of extremely low convergence of a solution because of a large fluctuation during calculation with an conventional optimization method using an inclination of the solution curved surface if the features of the solution space are not well known. On the other hand, in principle, the scale of the fluctuation parameter in our method has almost no influence on the process of obtaining an expected value. Although variations of convergence times of calculations due to the scale of the parameter are observed in actual calculations with the limited number of calculations, any large influence on the obtained expected values has not been known. Therefore, this method can be applied to unknown problems, including the surroundings, on solution spaces.

Also, the smaller the fluctuation parameter during calculation, the shorter the calculation time is, though this characteristic is not mentioned in this paper.

Restriction Conditions

In the conventional calculation approaches that start with initial solutions to search better performance index values, it may be difficult to generate an initial solution itself and thus to start the numerical calculation if the solution space is narrowed with restriction conditions. However, since the generation of solutions during calculation in our method is a stochastic process, it is possible to perform the calculation regardless of the presence of solutions.

Also, if there are restriction conditions on parameters, which are not the fluctuation parameter: h in (b) but are the load factor: n and the angle of attack: ⊠ etc. in Section 6.3 for example, it is necessary to keep parameters within the restriction conditions with general methods. However, with our method, such a device is not necessary. Furthermore, the restriction conditions are rather advantageous to numerical calculations. For example, if there are restrictions on parameter values and the range is specified, solutions can be generated only with values within the restrictions in the process where the series of solutions are generated. This is largely advantageous to perform numerical calculations. The harder the restriction conditions are and the narrower the range of the resultant parameter values is, the more densely the solutions are generated within the restriction conditions and the more easily the calculations are.

Next, in case of restriction conditions on state variables, when a solution generated during calculation is trapped on the restriction, the solution may be abandoned. However, expected values may approach but do not reach the restriction surfaces using only parameter values within constraint ranges and abandoning solutions that break the restrictions. Parameters and solutions that break the restrictions can be applied to a calculation using a penalty method to approach the expected value toward the restriction surface more closely. However, if results with this method are used as initial solutions of deterministic approaches, there is no necessity to make a device.

Scope of SPOT

The characteristics of SPOT are to obtain expected values of solution spaces with multiple integral. Thus, there is no problem if the solution spaces have restrictions as mentioned above. However, SPOT is not appropriate to be used for problems that have no restriction on the infinite solution spaces. Also, if a problem has a large solution space and a sharp peak on the performance index value, this method can be applied in principle. However, it may be difficult to obtain an approximate solution with high accuracy in the finite number of calculations. However, it is rare to have infinite range of a variable in an engineering problem, and the upper and lower limits can generally be set. Also, the fluctuation of a performance index value to that of a variable is expected to be smooth. Furthermore, the accuracy required for a variable is finite. Therefore, it is possible to obtain an approximate value for practical use with the finite number of calculations.

CONCLUSIONS

We propose a new optimization method based on the path integral called SPOT, and thus perform a numerical calculation on an engineering optimization problem. As a result, we obtain an approximate value of an optimal solution with high accuracy. In addition, this stochastic method has advantages on numerical calculation to other methods.

Also, SPOT has more effective features of optimization than the conventional methods on optimization problems of an unknown system or narrow solution space that leads to difficulty finding a feasible solution.

REFERENCES

1. Kirkpatrick, S., Gelatt Jr., C.D. and Vecchi, M.P. (1983) Optimization by Simulated Annealing. Science, 220, 671-680. http://dx.doi.org/10.1126/science.220.4598.671.
2. Terasaki, M., Kondo, M., Yoshida, H., Yamaguchi, K. and Ishikawa, Y. (2004) Integrated Optimization of Airplane Design and Flight Trajectory by New Optimization Method Using a Stochastic Process. In: CONPUTATIONAL MECHANICS WCCM VI in Conjunction with APCOM'04, Tsinghua University Press & Springer-Verlag, Beijing, 328.
3. Yoshida, H., Konno, T., Yamaguchi, K. and Ishikawa, Y. (2005) New Optimization Method Using Stochastic Process Based on Path Integral. Journal of the Japan Society for Computational Methods in Engineering, 5, 145-150. (in Japanese)
4. Yoshida, H., Yamaguchi, K. and Ishikawa, Y. (2006) Application of a New Optimization Method for System Design. Joint 3rd International Conference on Soft Computing and Intelligent Systems (SCIS) and 7th International Symposium on Advanced Intelligent Systems (ISIS), Tokyo, 20-24 September 2006, 1542-1547.
5. Yoshida, H., Yamaguchi, K. and Ishikawa, Y. (2007) Design Tool Using a New Optimization Method Based on a Stochastic Process. Transactions of the Japan Society for Aeronautical and Space Sciences, 49, 220-230. http://dx.doi.org/10.2322/tjsass.49.220
6. Feynman, R.P. and Hibbs, A.R. (1965) Quantum Mechanics and Path Integrals. McGraw-Hill, New York.
7. Metropolis, N., Rosenbluth, A.W., Rosenbluth, M.N., Teller, A.H. and Teller, E. (1953) Equation of State Calculations by Fast Computing Machines. Journal of Chemical Physics, 21, 1087-1092. http://dx.doi.org/10.1063/1.1699114

8. Nakane, M., Kobayashi, D., Yoshida, H. and Ishikawa, Y. (2009) Feasibility Study on Single Stage to Orbit Space Plane with RBCC Engine. 16th AIAA/DLR/DGLR International Space Planes and hypersonic Systems and Technologies Conference, Bremen, AIAA 2009-7331. http://dx.doi.org/10.2514/6.2009-7331
9. Ishimori, Y., Nakane, M., Ishikawa, Y., Yoshida, H. and Yamaguchi, K. (2011) Feasibility Study for Spaceplane's Concepts in Ascent Phase Using the Optimization Method. Journal of the Japan Society for Aeronautical and Space Sciences, 59, 291-297. (in Japanese)http://dx.doi.org/10.2322/jjsass.59.291
10. Suzuki, S. and Kawamura, N. (1996) Simultaneous Optimization of Sailplane Design and Its Flight Trajectory. Journal of Aircraft, 33, 567-571. http://dx.doi.org/10.2514/3.46982

CITATION

Yoshida, H, Yamaguchi, K. and Ishikawa, Y. (2014) Stochastic Process Optimization Technique. *Applied Mathematics*, 5, 3079-3090. doi: 10.4236/am.2014.519293.

Two Implicit Runge-Kutta Methods for Stochastic Differential Equation

Fuwen Lu, Zhiyong Wang*
Department of Mathematics, University of Electronic Science and Technology of China, Chengdu, China

ABSTRACT

In this paper, the Itô-Taylor expansion of stochastic differential equation is briefly introduced. The colored rooted tree theory is applied to derive strong order 1.0 implicit stochastic Runge-Kutta method (SRK). Two fully implicit schemes are presented and their stability qualities are discussed. And the numerical report illustrates the better numerical behavior.

INTRODUCTION

In this paper, we want to obtain numerical methods for strong solution of Stochastic Differential Equations of Itô type.

$$dy = f(y(t))dt + g(y(t))dW(t), y \in \mathbb{R} \tag{1.1}$$

Note that f is a slowly varying continuous component function, which is called drift coefficient, g is the rapidly varying continuous function called the diffusion coefficient. W(t) is a wiener process.

Recently, many scholars have successfully derived some methods for SDEs for both Itô and Stratonovich forms. Burrage and Burrage [1-3] established the colored rooted tree theory and Stochastic B-series expansion. Tian and Burrage [2,4,5] derived some strong order 1.0 2-stage Stochastic Runge-Kutta methods, including semiimplicit and implicit methods. Wang P. [6] derived some strong order 1.0 3-stage semi-implicit methods. Wang ZY [7] mainly considered the strong order SRKs for the SDEs of Itô form. In his PhD thesis he offered us the Colored Rooted tree theory for Itô tpye, and constructed some 2-stage and 3-stage explicit methods. Along this line, I will construct some implicit SRKs for SDEs of Itô type. In Section 2, the colored rooted tree theory for deriving SRK for SDEs of Itô type is briefly introduced and the 2 2-stage fully implicit SRKs are obtained. In Section 3 we will discuss their stability property. And in Section 4, we will report the numerical experiments.

2-STAGE IMPLICIT SRK AND ORDER CONDITIONS

Many scholars, including Burrage [2], offered the definition of the order of numerical methods in their thesis.

Definition 2.1: Let $\bar{y}_N$ be the numerical approximation to $y(t_n)$ after N steps with constant step size $(t_N - t_0)/N$; then $\bar{y}_N$ is said to be converge strongly to y with order p if

$$E\left(\left|\bar{y}_N - y(t_N)\right|\right) \leq Ch^p, \quad h \in (0, \delta) \tag{2.1}$$

Note that C is a constant that independent of h and $\delta > 0$.

Butcher presented the Rooted Tree theory, after which this theory was extended into stochastic area. Burrage [2] presented Colored Rooted Tree theory in her PhD thesis, and Wang [7] did the research especially for Itô SDEs. Similar to the deterministic condition, the definition of the elementary differential can be associated with $\forall t \in T$

$$F(\phi)(y) = y$$
$$F(\tau)(y) = f(y)$$
$$F(\delta)(y) = g(y)$$

$$F(t)(y) = f^{(m)}\left[F(t_1)(y),\cdots,F(t_m)(y)\right], t = [t_1,\cdots,t_m]$$
$$F(t)(y) = g^{(m)}\left[F(t_1)(y),\cdots,F(t_m)(y)\right], t = \{t_1,\cdots,t_m\}$$

Here ϕ stands for the trees having order 0.

Wang [7] deduced the Itô-Taylor series for SDEs. Firstly let's introduce two operators

$$L^0 = \frac{\partial}{\partial t} + f \cdot \frac{\partial}{\partial x} + \frac{1}{2} \cdot g^2 \cdot \frac{\partial^2}{\partial x^2}$$

$$L^1 = g \cdot \frac{\partial}{\partial x}$$

Now we introduce a very important proposition from Kloeden and Platen [8].

Proposition 2.1: if $A \subset M, h: \mathbb{R} \to \mathbb{R}$ is sufficiently derivative, and let X(t) be the solution of the equation

$$\begin{cases} dX(t) = f(X(t)) + g(X(t))dW(t), t > 0 \\ X(0) = X_0 \end{cases}$$

then

$$h(X(t)) = \sum_{a \in A} I_a\left[h_a(X_0)\right]_t + \sum_{a \in R(A)} I_a\left[h_a(X(\cdot))\right]_t \qquad (2.2)$$

Letting $h(X(t)) = X(t)$, then

$$X(t)$$
$$= X_0 + L^0 X_0 I_0 + L^1 X_0 I_1 + L^1 L^1 X_0 I_{11} + L^1 L^0 X_0 I_{10}$$
$$+ L^0 L^1 X_0 I_{01} + L^1 L^1 L^1 X_0 I_{111} + \cdots$$
$$= X_0 + fI_0 + gI_0 + g'gI_{11} + gfI_{10} + \left(fg' + \frac{1}{2}g^2 g''\right) I_{01}$$
$$+ g\left((g')^2 + gg''\right) I_{111} + \cdots$$

And from the definition of the elementary differential we can know

$$X(t) = F(\phi)(X_0) + F(\tau)(X_0)I_0 + F(\sigma)(X_0)I_1$$
$$\qquad + F(\{\sigma\})(X_0)I_{11} + F([\sigma])(X_0)I_{10}$$
$$\qquad + \left(F(\{\tau\})(X_0) + \frac{1}{2}F(\{\sigma,\sigma\})\right) I_{01}$$
$$\qquad + \left(F(\{\{\sigma\}\}) + F(\{\sigma,\sigma\})\right) I_{111}$$
$$\qquad + \cdots$$
$$= F(\phi)(X_0) + F(\tau)(X_0)I_0 + F(\sigma)(X_0)I_1$$
$$\qquad + F(\{\sigma\})(X_0)I_{11} + F([\sigma])(X_0)I_{10}$$
$$\qquad + F(\{\tau\})(X_0)I_{01} + F(\{\{\sigma\}\})(X_0)I_{111}$$
$$\qquad + F(\{\sigma,\sigma\})(X_0)\left(\frac{1}{2}I_{01} + I_{111}\right)$$
$$\qquad + \cdots$$
$$= \sum_{\rho(t) \leq 1.5} \alpha(t) F(t) I(t) + \cdots$$

Like the conclusion of Burrage [2], the Taylor-series of the actual solution of the SDEs is

$$X(t) = \sum_{t \in T} \alpha(t) F(t) I(t) \qquad (2.3)$$

The structure of Stratonovich-Taylor series is similar to the Itô-Taylor expansion, however, the stochastic calculations of these two types are different. Table 1 presents the trees and the corresponding elementary differentials. Especially, in order to illustrate the difference between Itô type and stratonovich type, we list all the stochastic calculations of trees having order ≤ 2.

Now we show general form of Runge-Kutta methods for SDEs of Itô form.

Let the stepsize of the methods is a constant $h = \dfrac{T}{N}, t_n = nh(n = 0,...,N)$, y_n is the numerical solution of $X(t)$, then

$$Y_i = y_n + \sum_{j=1}^{s} Z_{ij}^{(0)} \cdot f(Y_j) + \sum_{j=1}^{s} Z_{ij}^{(1)} \cdot g(Y_j)$$

$$y_{n+1} = y_n + \sum_{j=1}^{s} z_j^{(0)} \cdot f(Y_j) + \sum_{j=1}^{s} z_j^{(1)} \cdot g(Y_j)$$

(2.4)

Note that

$$Z_{ij}^{(0)} = h \cdot \alpha_{ij}, \quad i,j = 1,\cdots,s$$

$$z_j^{(0)} = h \cdot \alpha_j, \quad j = 1,\cdots,s$$

$$Z_{ij}^{(1)} = \sum_{l=1}^{p} b_{ij}^{(l)} \cdot \theta_l, \quad i,j = 1,\cdots,s$$

$$z_j^{(1)} = \sum_{l=1}^{p} \gamma_{ij}^{(l)} \cdot \theta_l, \quad j = 1,\cdots,s$$

where the $\theta_i (i = 1,\cdots,p)$ is random variables.

Using the Butcher Table, SRK can be written as

A	$B^{(1)}$	$B^{(2)}$	$\cdots$	$B^{(p)}$
α	$\gamma^{(1)}$	$\gamma^{(2)}$	$\cdots$	$\gamma^{(p)}$

Wang [7] deduced the Taylor series for the SRK of Itô form. And offered the definition of Elementary Weight, which has the same form of Burrage's conclusion [2].

Definition 2.2:

$$\Phi(t) = \begin{cases} e, & t = \varnothing \\ l(t) \cdot z^{(0)T} \cdot \prod_{i=1}^{\lambda} \Psi(t_i), & t = [t_1, \cdots, t_\lambda] \\ l(t) \cdot z^{(1)T} \cdot \prod_{i=1}^{\lambda} \Psi(t_i), & t = \{t_1, \cdots, t_\lambda\} \end{cases}$$

where

$$\Psi(t) = \begin{cases} l(t) \cdot z^{(0)T} \cdot \prod_{i=1}^{\lambda} \Psi(t_i), & t = [t_1, \cdots, t_\lambda] \\ l(t) \cdot z^{(1)T} \cdot \prod_{i=1}^{\lambda} \Psi(t_i), & t = \{t_1, \cdots, t_\lambda\} \end{cases}$$

As the definition of Elementary Weight that we obtained, we can gain the stochastic Runge-Kutta series expansion

$$Y(t) = \sum_{t \in T} \frac{\alpha(t) \cdot \Phi(t) \cdot F(t)(y(t_0))}{l(t)!} \tag{2.5}$$

Table 2 offers the trees and their Elementary Weights.

From the Equations (2.4) and (2.5) we can obtain the truncation error at $t=t_n$.

Two Implicit Runge-Kutta Methods for Stochastic Differential

Table 1: Trees and the corresponding elementary differentials.

$\rho(t)$	t	I(t)	$\rho(t)$	t	I(t)
0	ϕ	1	2	$\{\sigma,\tau\}$	I_{011}
0.5	σ	I_1	2	$\{\tau,\sigma\}$	I_{101}
1	τ	I_0	2	$\{\{\tau\}\}$	I_{011}
1	$\{\sigma\}$	I_{11}	2	$\{\sigma,\sigma,\sigma\}$	$I_{111} + \frac{1}{2}I_{011} + \frac{1}{2}I_{101}$
1.5	$[\sigma]$	I_{10}	2	$\{\{\sigma\},\sigma\}$	$2I_{111} + \frac{1}{2}I_{011} + \frac{1}{2}I_{01}$
1.5	$\{\tau\}$	I_{10}	2	$\{[\sigma]\}$	I_{101}
1.5	$\{\{\sigma\}\}$	I_{111}	2	$[\{\sigma\}]$	I_{110}
1.5	$\{\sigma,\sigma\}$	$\frac{1}{2}I_{01} + I_{111}$	2	$\{\sigma\{\sigma\}\}$	$I_{1111} + \frac{1}{2}I_{101}$
2	$[\tau]$	I_{00}	2	$\{\{\{\sigma\}\}\}$	$I_{1111} + \frac{1}{2}I_{011}$

Table 2: Trees and the corresponding elementary weights.

$\rho(t)$	t	$\phi(t)$	$\rho(t)$	t	$\phi(t)$
0	ϕ	e	1.5	$[\sigma]$	$2z^{(0)T}Z^{(1)}e$
0.5	σ	$Z^{(1)T}e$	1.5	$\{\tau\}$	$2z^{(1)T}Z^{(0)}e$
1	τ	$2z^{(0)T}e$	1.5	$\{\{\sigma\}\}$	$6z^{(1)T}Z^{(1)}Z^{(1)}e$
1	$\{\sigma\}$	$2z^{(1)T}Z^{(1)}e$	1.5	$\{\sigma,\sigma\}$	$3z^{(1)T}\left(Z^{(1)}e\right)^2$

$$L_n = \sum_{t \in I}\left(I(t) - \frac{\Phi((t))}{I(t)!}\right)\alpha(t)F(t)(y(t_n))$$

$$= \sum_{t \in T} e(t)\alpha(t)F(t)(y(t_n))$$

Proposition 2.2, given by Burrage and Burrage [3], gives the necessary conditions of the methods.

Proposition 2.2: L_n is the local truncation error of the numerical methods at $t = t_n$, ε_N is the global error at $t = t_N$, if f and g is sufficiently derivative, and $\forall n = 1, \cdots, N$

$$\left(E\left[\|L_n\|^2\right]\right)^{\frac{1}{2}} = O\left(h^{p+\frac{1}{2}}\right)$$

$$E[L_n] = O\left(h^{p+1}\right)$$

then

$$E[\varepsilon_N] = O\left(h^p\right)$$

From the Proposition 2.2, the Runge-Kutta methods of the strong order 1.0 have to satisfy

1) $\forall t$ that $p(t) \leq 1 \Leftrightarrow \left(E\left[(e(t))^2\right]\right)^{\frac{1}{2}} = 0$
$$E\left[(e(t))^2\right] = 0 \qquad (2.6)$$

2) $\forall t$ that $p(t) \leq 1.5 \quad E[e(t)] = 0$

obviously, $\forall t, E\left[(e(t))^2\right] = 0 \Rightarrow E[e(t)] = 0$, thus in 2) We just need to consider the condition when $p(t) = 1.5$.

Two Implicit Runge-Kutta Methods for Stochastic Differential

Now we introduce the random variables $\theta_1 = I_1, \theta_2 = \sqrt{h}$. And we note $c = A \cdot e, b = B^{(1)} \cdot e, d = B^{(2)} \cdot e, \lambda = b \cdot I_1 + d \cdot \sqrt{h}$

Now let's start to construct the methods of strong order 1.0.

1) For tree σ

$$E\left[\left(I_1 - z^{(1)T}e\right)^2\right]$$
$$= E\left[\left(I_1 \cdot (1 - \gamma^{(1)T}e) - \sqrt{h} \cdot \gamma^{(2)T}e\right)^2\right] = 0$$

namely

$$\left(1 - \gamma^{(1)T}e\right)^2 \cdot \frac{h^2}{2} + \left(\gamma^{(2)T}e\right)^2 \cdot h = 0$$
$$\Rightarrow \gamma^{(1)T}e = 1, \gamma^{(2)T}e = 0$$

2) For tree τ

$$E\left[\left(I_0 - z^{(0)T}e\right)^2\right] = \left(1 - \alpha^T e\right)^2 \cdot h^2 = 0$$
$$\Rightarrow \alpha^T e = 1$$

3) For tree $\{\sigma\}$

$$E\left[\left(I_{11} - z^{(1)T}Z^{(1)}e\right)^2\right] = 0$$

namely

$$E\left[\left(I_1^2\cdot\left(\frac{1}{2}-\gamma^{(1)T}b\right)-I_1\cdot\left(\gamma^{(1)T}d+\gamma^{(2)T}b\right)\right.\right.$$
$$\left.\left.-h\cdot\left(\frac{1}{2}+\gamma^{(2)T}d\right)\right)^2\right]$$
$$=0$$
$$\Rightarrow X^T DX = 0$$

where

$$X=\left(\frac{1}{2}-\gamma^{(1)T}b,\gamma^{(1)T}d+\gamma^{(2)T}b,\frac{1}{2}+\gamma^{(2)T}d\right)$$

and

$$D=\begin{pmatrix}3 & 0 & 1\\ 0 & 1 & 0\\ 1 & 0 & 1\end{pmatrix}$$
$$\Rightarrow X=0$$

namely

$$\gamma^{(1)T}b=\frac{1}{2},\gamma^{(1)T}d+\gamma^{(2)T}b=0,\gamma^{(2)T}d=-\frac{1}{2}$$

4) For tree $[\sigma]$

$$E\left[I_{10}-z^{(0)T}Z^{(1)}e\right]=E\left[I_{10}-\alpha^T\cdot h\left(b\cdot I_1+d\cdot\sqrt{h}\right)\right]=0$$
$$\Rightarrow \alpha^T\cdot d=0$$

5) For tree $\{\tau\}$

Two Implicit Runge-Kutta Methods for Stochastic Differential

$$E\left[I_{01} - z^{(1)T} Z^{(0)} e\right]$$
$$= E\left[I_{01} - h \cdot \left(\gamma^{(1)T} \cdot c \cdot I_1 + \gamma^{(2)T} \cdot c \cdot \sqrt{h}\right)\right] = 0$$
$$\Rightarrow \gamma^{(2)T} \cdot c = 0$$

6) For tree $\{\{\sigma\}\}$

$$E\left[I_{111} - \left(\gamma^{(1)T} I_1 + \gamma^{(2)T} \sqrt{h}\right)\left(B^{(1)} I_1 + B^{(2)} \sqrt{h}\right)\left(bI_1 + d\sqrt{h}\right)\right]$$
$$= 0$$
$$\Rightarrow \gamma^{(1)T} B^{(1)} d + \gamma^{(1)} B^{(2)} b + \gamma^{(2)} B^{(1)} b + \gamma^{(2)} B^{(2)} d = 0$$

7) For tree $\{\sigma, \sigma\}$

$$E\left[\frac{1}{2} I_{01} + I_{111} - \frac{1}{2} z^{(1)T} \cdot \left(Z^{(1)} e\right)^2\right] = 0$$
$$\Rightarrow 2\gamma^{(1)T} bd + \gamma^{(2)T} b^2 + \gamma^{(2)T} d^2 = 0$$

Thus, the 2-stage implicit SRKs should satisfy the system

$$\begin{cases} \gamma^{(1)T} e = 1 \\ \gamma^{(2)T} e = 0 \\ \alpha^T e = 1 \\ \gamma^{(1)T} b = \frac{1}{2} \\ \gamma^{(1)T} d + \gamma^{(2)T} b = 0 \\ \gamma^{(2)T} d = -\frac{1}{2} \\ \alpha^T \cdot d = 0 \\ \gamma^{(2)T} \cdot c = 0 \\ \gamma^{(1)T} B^{(1)} d + \gamma^{(1)} B^{(2)} b + \gamma^{(2)} B^{(1)} b + \gamma^{(2)} B^{(2)} d = 0 \\ 2\gamma^{(1)T} bd + \gamma^{(2)T} b^2 + \gamma^{(2)T} d^2 = 0 \end{cases} \quad (2.7)$$

Here we gained the conditions for the methods with strong order 1.0, theoretically we can construct any-stage methods, both explicit and implicit. And now we consider the 2-stage implicit methods.

$$\begin{array}{cc|cc|cc} a_{11} & 0 & b_{11} & b_{12} & d_{11} & d_{12} \\ 0 & a_{22} & b_{21} & b_{22} & d_{21} & d_{22} \\ \hline \alpha_1 & \alpha_2 & \gamma_1^{(1)} & \gamma_2^{(1)} & \gamma_1^{(2)} & \gamma_2^{(2)} \end{array}$$

Bringing the table into the system 2.7, and letting the

, $a_{11} = a_{22} = \dfrac{1}{2}, \alpha_1 = 1, \alpha_2 = 0$, we can obtain the first scheme—Imp$_1$

Imp$_1$

$$\begin{array}{cc|cc|cc} \dfrac{1}{2} & 0 & \dfrac{1}{3} & \dfrac{1}{3} & -\dfrac{1}{3} & \dfrac{1}{3} \\ 0 & \dfrac{1}{2} & \dfrac{1}{2} & \dfrac{1}{2} & 1 & 0 \\ \hline 1 & 0 & \dfrac{1}{2} & \dfrac{1}{2} & \dfrac{1}{2} & -\dfrac{1}{2} \end{array}$$

Furthermore if we continue to let $b_{11} = b_{22} = d_{11} = d_{22} = 0$, we can obtain another scheme —Imp$_2$.

Imp$_2$

$$\begin{array}{cc|cc|cc} \dfrac{1}{2} & 0 & 0 & 0 & 0 & 0 \\ 0 & \dfrac{1}{2} & 0 & 1 & 1 & 0 \\ \hline 1 & 0 & \dfrac{1}{2} & \dfrac{1}{2} & \dfrac{1}{2} & -\dfrac{1}{2} \end{array}$$

STABILITY

Saito and Mitzui [9] introduced the definition of meansquare(MS) stability, and the scholars such as Burrage [2] and Tian [4,5] researched it and gave some improvements.

Consider the linear test equation of Itô type of SDEs.

$$dy = \lambda y dt + \mu y dw(t) \tag{3.1}$$

and we use one-step scheme

$$y_{n+1} = R(h, \lambda, \mu, I) y_n$$

where h is the stepsize, I is the random variable in the numerical scheme.

Satio and Mitzui [9] introduced the definition Definition 3.1. If for λ, μ, h,

$$\bar{R}(h, \lambda, \mu) = E\left(R^2(h, \lambda, \mu, I)\right) < 1$$

then the numerical scheme is said to be MS stable, and the $\bar{R}(\lambda, \mu, h)$ is said to be the MS-stability function.

1) For Imp_1, we can obtain the MS-stability function

$$y_{n+1} = R(h, \lambda, \mu, I_{n1}) y_n$$

where

$$R(h, \lambda, \mu, I_{n1}) = 1 + R_1 \cdot p$$
$$+ \frac{1}{2}(R_1 + R_2) \cdot q \cdot I_{n1} + \frac{1}{2} \cdot (R_1 - R_2) \cdot q$$

Note that

$$R_1 = -2(5q \cdot I_{n1} + 3p - 2q - 6) R_3$$

$$R_2 = -2(-q \cdot I_{n1} - 6 - 8q + 3p) R_3$$

$$R_3 = \frac{1}{4q^2 I_{n1}^2 - 4q^2 + 12 - 12p - 10q \cdot I_{n1} + 3p^2 + 5pq \cdot I_{n1} + 4p - 2pq}$$

and

$p = \lambda h$, $q = \mu\sqrt{h}$, I_{N1} is the is the standard Gaussian variable $\sim N(0,1)$
Figure 1 describes the stable region of Imp_1.

2) For the method Imp_2, we obtain that

$$y_{n+1} = R(h, \lambda, \mu, I_{n1}) y_n$$

where

$$R(h, \lambda, \mu, I_{n1}) = 1 + pR_1$$

$$+ q\left(R_1\left(\frac{1}{2}I_{n1} + \frac{1}{2}\right) + R2\left(\frac{1}{2}I_{n1} - \frac{1}{2}\right) \right)$$

Note that

$$R_1 = \frac{1}{1 - \frac{1}{2}p}$$

$$R_2 = \frac{1+qR_1}{1-\frac{1}{2}p-qI_{n1}}$$

and

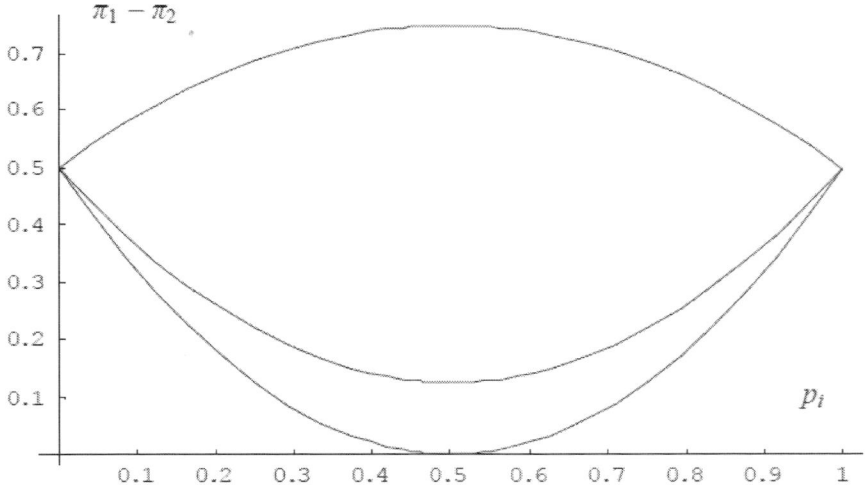

Figure 1: Stable region of Imp$_1$.

$p = \lambda h$, $q = \mu\sqrt{h}$, I_{N1} is the is the standard Gaussian variable $\sim N(0,1)$
Figure 2 represents the stable region of Imp$_2$.

NUMERICAL RESULTS

Now we report the numerical results of the schemes derived in this paper. At first we will use the points of numerical simulation in a single trajectory to compare the absolute error Ms of five different schemes explicit Euler-Maruyama scheme, explicit milstein scheme, explicit

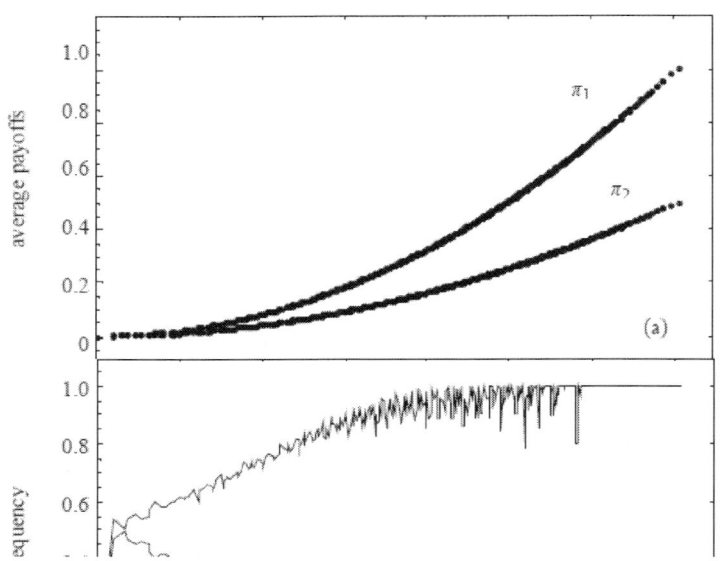

Figure 2: Stable region of Imp$_2$.

two-stage scheme I$_{21}$ which is designed by Wang [7], Imp$_1$ and Imp$_2$—for a same non-linear system 10. After which we will simulate 100 trajectories of each scheme and then compare their absolute error Ms.

Errors for the (4.1) is given by

$$M = \frac{1}{k}\sum_{i=1}^{k}|x_i - y(t_i)|$$

Note that x_i is the exact value at step point t_i and $y(t_i)$ is the numerical simulation at that point, k is the number of the points chosen in the trajectories. And the non-linear system (4.1) is given by

$$\begin{cases} dX(t) = \left(\frac{1}{2}X(t) + \sqrt{X^2(t)+1}\right) \cdot dt \\ \qquad + \sqrt{X^2(t)+1} \cdot dw(t), \ t \in [0,5] \\ X(0) = 0 \end{cases}$$

(4.1)

And the analytical solution of the system 10 is

$$X(t) = \sinh(t + w(t)) \qquad (4.2)$$

Firstly, we compare the error Ms in a single trajectory. From the Table 3, we can know that in a random trajectory(actually we choose the first one), the Imp_1 is obviously better than all the other schemes, and also,

Table 3: The absolute error Ms in a single trajectory

stepsize	2^{-4}	2^{-5}	2^{-6}	2^{-7}	2^{-8}
Euler	12.97	7.47	1.87	0.98	5.38
milstein	14.75	7.24	3.44	1.90	8.68
I_{21}	14.68	7.19	3.37	1.85	8.73
Imp_1	1.86	1.52	0.34	0.15	2.72
Imp_2	38.01	13.63	4.30	2.16	5.38

Table 4: Mean of the absolute error Ms in 100 trajectories

stepsize	2^{-4}	2^{-5}	2^{-6}	2^{-7}	2^{-8}
Euler	12.97	7.47	1.87	0.98	5.38
milstein	14.75	7.24	3.44	1.90	8.68
I_{21}	14.68	7.19	3.37	1.85	8.73
Imp_1	1.86	1.52	0.34	0.15	2.72
Imp_2	38.01	13.63	4.30	2.16	5.38

Imp_2 has a same accuracy with I_{21} scheme and milstein scheme.

Now let's contrast the absolute error Ms of 100 trajectories.

From the Table 4, we can conclude that Imp_1 is obviously better than all the other schemes, especially when $h = 2^{-4}, 2^{-6}, 2^{-7}$. Still, Imp_2 always has a same accuracy with I_{21} scheme and milstein scheme. It shows that Imp_1 is better than other schemes, and Imp_2 is also a proper scheme for solving stochastic differential equations.

REFERENCES

1. K. Burrage and P. M. Burrage, "High Strong Order Explicit Runge-Kutta Methods for Stochastic Ordinary Differential Equations," Applied Numerical Mathematics, Vol. 22, 1996, pp. 81-101. doi:10.1016/S0168-9274(96)00027-X
2. P. M. Burrage, "Runge-Kutta Methods for Stochastic Differential Equations," Ph.D. Thesis, The University of Queensland, Queensland, 1999.
3. K. Burrage and P. M. Burrage, "Order Condition of Stochastic Runge-Kutta Methods by B-Series," SIAM Journal on Numerical Analysis, Vol. 38, No. 5, 2000, pp. 1626-1646.doi:10.1137/S0036142999363206
4. T. H. Tian, "Implicit Numerical Methods for Stiff Stochastic Differential Equations and Numerical Simulations of Stochasic Models," Ph.D. Thesis, The University of Queensland, Queensland, 2001.
5. T. H. Tian and K. Burrage, "Two Stage Runge-Kutta Methods for Stochastic Differential Equations," BIT, Vol. 42, No. 3, 2002, pp. 625-643. doi:10.1023/A:1021963316988
6. P. Wang, "Three-Stage Stochastic Runge-Kutta Methods for Stochastic Differential Equaitons," Journal of Computational and Applied Mathematics, Vol. 222, No. 2, 2008, pp. 324-332. doi:10.1016/j.cam.2007.11.001
7. Z. Y. Wang, "The Stable Study of Stochastic Functional Differential Equation," Ph.D. Theis, Huazhong University of Science and Technology, Wuhan, 2008
8. P. E. Kloeden and E. Platen, "Numerical Solution of Stochastic Differential Equations," Springer-Verlag, Belin, 1992.
9. Y. Saito and T. Mitsui, "Stability Analysis of Numerical Schemes for Stochastic Differential Equations," SIAM Journal on Numerical Analysis, Vol. 33, No. 6, 1996, pp. 2254-2267.doi:10.1137/S0036142992228409

CITATION

F. Lu and Z. Wang, "Two Implicit Runge-Kutta Methods for Stochastic Differential Equation," *Applied Mathematics*, Vol. 3 No. 10, 2012, pp. 1103-1108. doi: 10.4236/am.2012.310162.

Parameter Dependence in Stochastic Modeling—Multivariate Distributions

Jerzy K. Filus[1] and Lidia Z. Filus[2]
[1]Department of Mathematics and Computer Science, Oakton Community College, Des Plaines, USA
[2]Department of Mathematics, Northeastern Illinois University, Chicago, USA

ABSTRACT

We start with analyzing stochastic dependence in a classic bivariate normal density framework. We focus on the way the conditional density of one of the random variables depends on realizations of the other. In the bivariate normal case this dependence takes the form of a parameter (here the "expected value") of one probability density depending continuously (here linearly) on realizations of the other random variable. The point is, that such a pattern does not need to be restricted to that classical case of the bivariate normal. We show that this paradigm can be generalized and viewed in ways that allows one to extend it far beyond the bivariate or multivariate normal probability distributions class.

INTRODUCTION

This paper can be viewed as an extension of our previous work (Filus and Filus [1]) on the bivariate Gaussian pdf structure's genesis of wide classes of newly constructed bivariate probability distributions. These distributions we constructed in our papers since 2000 up to recently (see, Filus and Filus [1] -[6] , also see Kotz, Balakrishnan and Johnson [7] pp. 217-218). Here, our explanations concerning the relation of our models to the bivariate normal were modified and we extended the topic to higher dimensional models and to the relation between our "parameter dependence method" of models construction and the multivariate normal paradigm.

It is a well-known fact that among existing multivariate probability distributions, there are no more than a few classes that are widely and successfully applied in practical stochastic modeling procedures. Typically, the underlying random variables are assumed to be independent or having an "approximately Gaussian" bivariate or multivariate distribution. The normality often is assumed even when corresponding data hardly agree with that mathematical model (showing asymmetry, for example). On the other hand, from all the multivariate distributions used in applications, the normal seems to be "the best". The reason for this is that the Gaussian models catch the stochastic relationship (mainly by a regression function) between its marginal random variables in the most natural way. We first analyze and interpret the specific way the multivariate normal density of the random vector $(X_1,...,X_m)$ relates to the marginal quantities $X_1,...,X_m$. Next we extend the "Gaussian pattern" to more general classes of bivariate and multivariate distributions including cases with non-Gaussian marginals. First of all we show that the common "mechanism" of the stochastic dependences both in the Gaussian distributions structure and the structure of the distributions we define, relies on the same way of conditioning. Namely, in all the considered cases, the conditional density of one random variable, say, $X_j (j=2,3,...,m)$, given realizations, say, $X_1,...,X_{j-1}$, of the marginal random variables $X_1,...,X_{j-1}$ can be obtained by setting (often arbitrarily) an "initially constant" parameter, say, q_j of the density of X_j to be any continuous function $\theta_j(x_1,...,x_{j-1})$ of the realizations. In this way the conditional pdf of X_j, given $x_1,...,x_{j-1}$ is defined, which stands for the "source" of the stochastic dependences.

Pursuing this method successively for $j=2,3,...,m$ we always arrive at a unique model. This manner, however, is characteristic for the m-variate normal where we turn an "original" normal $N(m_j, \sigma_j)$ density of X_j into the conditional density by setting a new value μ_j^* of the "affected

Parameter Dependence in Stochastic Modeling—Multivariate

old parameter m_j" to be the following (linear regression) function: $\mu_j^* = \mu_j + c_1 x_1 + \ldots + c_{j-1} x_{j-1}$. Our main contribution is to generalize the latter function to any, not necessarily linear, function and consider not only the parameter m_j but also any other parameter of any probability density to define the corresponding conditional distributions. This is the essence of the socalled parameter dependence method. Specifically in this paper, our task will be showing more closely relation of this method to the multivariate Gaussian model construction.

In Section 2, we analyze the stochastic dependences between marginal random variables of the bivariate normal in order to point out the original version of the parameter dependence pattern next extending to other constructed bivariate probability densities. The explanation as well as the example of applications of the bivariate normal is different from that in Filus and Filus [1]. In Section 3 we present the extension of the bivariate normal pdf to the bivariate FF-normal (formerly called "pseudo-normal"). In Section 4 we apply the parameter dependence method to construct the bivariate FF-Weibull (formerly "pseudo-Weibull") density in reliability framework of joint density of parallel system components life times. Comparison with other, similar, methods in the literature is presented in Section 5. In Section 6 we extend the constructions from bivariate to multivariate probability densities, first showing their relation with the multivariate normal dependences structure. Examples of the construction of multivariate FF-normal and multivariate FF-exponential ("pseudo-exponential") densities are given.

In Section 7, we point out that the "method of parameter dependence" is used in some more areas of reliability theory for different situations than we are considering. This is a part of the accelerated life testing theory where the dependence of life time distribution's parameter from a given (high) stress is investigated.

Another (fairly new) area is the "load optimization theory" sometimes associated with the load sharing phenomena analysis that we sketch in Subsection 7.2. The differences between these approaches and our theory are pointed out in 7.2.

THE BIVARIATE NORMAL CASE

We start with the following situation. Suppose the normally distributed random variable X_2 describes an attribute of a physical or biological object, say u. Consider the (stochastic) behavior of the object u in two distinct "physical" situations. In the first situation, u is exposed to some random stress whose magnitude is described by a normally distributed random variable X_1. In the second situation we assume there is no such a stress present or the stress takes on a fixed predetermined value. The usual task here is to determine the joint distribution of X_1, X_2. Let the densities of X_1, X_2 be normal, i.e., $g_1(x_1)=N(\mu_1, \sigma_1)$ $g_2(x_2)=N(\mu_2, \sigma_2)$. [It is clear that we must assume that the value $(m_2 - k\sigma_2)$ is positive for at least $k = 3$, in order to assure approximate positivity of the normal life-time X_2].

Simplified Biomedical Example

Imagine the following fictitious experiment whose goal is to establish the possible stochastic impact of a medication's dose change on some cancer treatment results. Suppose a person of a certain fixed age, was diagnosed with a kind of cancer. Assume that one of the significant characteristics of that kind of cancer is a tumor with a size X_2. During a given time period T after the patient was diagnosed, a specific medication was administered. Also suppose that this medication was routinely administered in the past, and that the average dose is estimated (or fixed) to be m_1 milligrams per kilo of weight daily. Assume that, originally, the known (either measured or estimated) average size of the tumor is, say, m_2 millimeters and after the period T of treatment the tumor size X_2 is measured again and its negative or positive increment $X_2-\mu_2$ is statistcally confronted with the dose X_1 of the medication administered.

We assume that the goal of the underlying experiment is to make a prediction on effect $X_2-\mu_2$ of the treatment when the dose is changed from its "original level" $X_1-\mu_1$ to a level $X_1-\mu_1$. Randomness of the doze X_1 may be justified when only "historical data" are analyzed and then extrapolated for a larger population of cases not yet recorded. In the case of extrapolation of historical data for a larger population we assume

Parameter Dependence in Stochastic Modeling—Multivariate

that the only information one possess on the applied in the past dose X_1 is its probability distribution, which is the Gaussian $N(\mu_1, \sigma_1)$ with given values for both the parameters m_1, σ_1. Also, for X_1-μ_1 the tumor size X_2 (after the treatment) is assumed to be a random variable having a normal $N(\mu_2, \sigma_2)$ density, where X_2-μ_2 is the value of the tumor "increment" under the treatment characterized by the dose level m_1. For any other applied doze $X_1 = x_1$ the, associated with a single patient, value $(x_1 - m_1)$ statistically affects the change in the tumor size $X_2 - m_2$ i.e., the treatment result. The word "statistically" here means that the impact of a nonzero quantity $(x_1 - m_1)$ ("the dose is not the standard one") on the (former) probability density $N(m_2, \sigma_2)$ of the tumor size X_2 realizes through affecting the value of the mean m_2 rather than directly affecting the numerical value x_2 of X_2.

If we were interested in finding the joint probability distribution of X_1, X_2 it is enough to determine the conditional density $g_2(x_2|x_1)$ of $X_2|X_1$, since the marginal density of X_1 is not changing.

In accordance with the "linear regression rule", the dependence of the (new) expected value μ_2^* (so new probability density $N(\mu_2^*, \sigma_2^*)$) of the tumor size X_2 on the event $X_1 = x_1$, is determined by the familiar functional relationship:

$$E(X_2|X_1 = x_1) = \mu_2^* = \mu_2 + a(x_1 - \mu_1) \tag{1}$$

where $a = r(\sigma_2/\sigma_1)$ and r is the (linear) correlation coefficient of the variables X_1, X_2.

This approach directly leads to the determination of the conditional density of the random variable X_2 given any realization $X_1 = x_1$. It is a well-known fact that the conditional density $g_2(x_2|x_1)$ is, again, normal and

$$g_2(x_2|x_1) = \frac{1}{\sigma_2\sqrt{2\pi(1-\rho^2)}} \exp\frac{-(x_2 - \mu_2 - a(x_1 - \mu_1))^2}{2\sigma_2^2(1-\rho^2)}, \tag{2}$$

i.e., the $N\left(\mu_2, a(x_1-\mu_1), \sigma_2\sqrt{1-\rho^2}\right)$ density in x_2.

The joint density $g(x_1,x_2)$ of the random variables X_1, X_2 is given by the usual arithmetic product $g_2(x_2|x_1)g_1(x_1)$.

Remark

In the example above, one can reinterpret the "response random variable" X_2 to be for example the patient's "residual life-time", or blood pressure, or level of some important chemical in the blood (such as cholesterol). In such cases the mathematics of the problem would remain the same.

Note the obvious fact that the tumor size X_2 does not have a physical influence on the medication dose X_1 so that the original marginal pdf $g_1(x_1)$ remains the same. However, the stochastic dependence between X_1, X_2 is mutual, since, in general, $g_1(x_1|x_2) \neq g_1(x_1)$.

It is well known that the actual problem with the bivariate normal density construction is to get to the conditional density (2), which fully represents the underlying stochastic dependence of random variable X_2 on X_1.

Our claim is that the above paradigm for the stochastic dependence (characteristic for the bivariate Gaussians) can be extended to other classes of bivariate and multivariate distributions (see Filus and Filus [3] [4]).

THE FF-NORMAL (PSEUDONORMAL) EXTENSION

Historically, people relied on the nice symmetry in the stochastic dependence of X_1 and X_2 when using their joint bivariate normal distribution. This kind of symmetry (i.e., both marginal and both conditional distributions are normal and both sides regression functions are linear) can only be achieved with the linear regression functions as described above (1). However, are linear regression functions really the

Parameter Dependence in Stochastic Modeling—Multivariate

only functions that one can successfully apply within this framework? Assuming that the function $m_2(X_1)$ is any continuous function in X_1, one obtains a wide and interesting extension of the class of bivariate normal densities. We called this class FF-normal (previously named "pseudonormal", see [2] and also [7]). In this case the parameter σ_2 can as well become a continuous function of the stress X_1 (X_1 may have a "stress" interpretation in a very wide sense). This stress may change the parameter σ_2 of the (normal) density of, say, the "life-time" X_2 into another value $\sigma_2^* = \sigma_2^*(x_1)$ that depends on the particular realization x_1 of the random variable X_1. The price for such a wide generalization is loss of the, mentioned above, symmetry (the marginal of X_2 ceases to be normal) but the gains are considerable. Anyway, the bivariate normal remains to be a special cases of the FF-normals.

With the bivariate FF-normal densities of (X_1, X_2) we can use general continuous $m_2(x_1)$ and $\sigma_2(x_1)$ functions, and, performing similar calculations as above, we find, rather surprisingly, that $g_2(x_2|x_1)$ is once more a regular normal density in x_2.

Consider now the following situation with a bivariate FF-normal distribution in which the "physical" interpretation of the underlying random variables can now be more general than above. Let u_1, u_2 be two objects (or phenomena) which are characterized by the random variables X_1, X_2 respectively. If the objects are physically separated then the random variables X_1, X_2 are assumed to be independent, having normal pdfs $g_1(x_1)=N(\mu_1, \sigma_1)$ and $g_2(x_2)=N(\mu_2, \sigma_2)$ respectively. When the objects physically interact (or rather only u_1 physically impacts u_2), then the corresponding joint FF-normal density $g(x_1, x_2)$ of the random vector (X_1, X_2) is given by the usual product formula:

$g(x_1,x_2)=g_1(x_1)g_2(x_2|x_1)$ with the invariant marginal density $g_1(x_1)=N(\mu_1, \sigma_1)$.

For the conditional density of $X_2|x_1$ we have: $g_2(x_2|x_1) = N(\mu_2^*(x_1), \sigma_2^*(x_1))$.

The functions $\mu_2^*(x_1), \sigma_2^*(x_1)$ are "formed" from the "no-stress" parameters m_2, σ_2 of X_2's density.

More explicitly, one obtains the bivariate FF-normal pdf in the form:

$$g(x_1,x_2) = \frac{1}{\sqrt{2\pi}\sigma_1} \exp\frac{-(x_1-\mu_1)^2}{2\sigma_1^2} \frac{1}{\sqrt{2\pi}\sigma_2^*(x_1)} \exp\frac{-(x_2-\mu_2^*(x_1))^2}{2\sigma_2^{*2}(x_1)}$$

where $\mu_2(x_1) = E[X_2|x_1]$ is the (in general, nonlinear) "regression function", and $[\sigma_2^*(x_1)]^2 = \text{Var}[X_2|x_1]$ is the conditional variance obtained from the "previous" variance σ_2^2.

In particular, one may consider the "nonlinear regression function"

$$E[X_2|x_1] = \mu_2(x_1) = \mu_2 + a^*(x_1-\mu_1) + A(x_1-\mu_1)^n$$

with arbitrary real parameters a^* and A, $n=2,3,...$

Realize that in the case $A = 0$ and $\sigma_2(x_1) = \sigma_2\sqrt{1-\rho^2}$ we obtain the regular bivariate Gaussian density. The coefficient A of the term $A(x_1-\mu_1)^n$ may be considered as a nonlinear "correction" of the regular Gaussian (linear) regression. The main purpose of that correction is to enhance the accuracy in various modeling procedures. For some type of asymmetric data, especially interesting may be the "non-symmetric" "quadratic" case $n = 2$.

Equally important is the "symmetric" "cubic" case when $n = 3$. Also interesting is the case where $a^* = 0$ and $A \neq 0$. Here, the linear correlation coefficient is zero, while the (nonlinear) dependence may be quite "heavy". Another interesting case (which may be combined with those above) is when we choose the standard deviation function as belonging to the "parametric class" of "quadratic functions" defined

Parameter Dependence in Stochastic Modeling—Multivariate

by $\sigma_2^*(x_1) = k\sigma_2(1+r^2x_1^2)$ with k positive real. It is rather obvious that the idea of the construction of bivariate normal (and FF-normal as well) can be extended to other probability distributions such as exponential, gamma, Weibull, lognormal, etc.

Generally speaking, the essence of the construction method is that for any pair of ("initially independent") random variables X_1 and X_2 with given probability densities $g_1(x_1, \omega)$ and $g_2(x_2; \Theta)$ respectively, one can simply "declare" some parameter (or a vector parameter), say Θ, of the density $g_2(x_2; \Theta)$ to be dependent on the values of the other random variable X_1. This means that when $X_1 = x_1$ we may assume that $g_2(x_2; \Theta(x_1)) = g_2(x_2|x_1)$ is the "affected by x_1" distribution of the random variable X_2 (given the event $X_1 = x_1$ occurred, with probability density $g_1(x_1)$).

Then the joint density of the pair (X_1, X_2) is always

$$g(x_1, x_2) = g_2(x_2, \Theta(x_1))g_1(x_1).$$

This situation is especially natural if we consider X_2 to be the life-time of an object and X_1 is the stress put on it.

Roughly, one can say that the construction method of bivariate distributions, presented above is an extension of the method used in the construction of the bivariate normal.

RELIABILITY EXAMPLE

Consider a 2-component (say u_1, u_2) parallel system reliability setting in which X_1, X_2 represent the components' life-times (see Barlow and Proschan [8]). We start with the situation where the system's components act separately. We call that pattern the "laboratory conditions". In this latter case the components are physically separated and consequently their life-times (represented by the, statistically estimated, "baseline probability densities" $g_1(x_1)$ and $g_2(x_2)$ respectively), are stochastically independent. When the two components are installed into the system, they start to interact. Assume that during that interaction some irregularities in the work of component u_1 cause corresponding changes in u_2's inner physical structure. This increases the hazard rate of that physically affected component u_2. Such physical phenomena are then "responsible" for the occurrence of stochastic dependence in the "in-system" component life-times X_1 and X_2.

One can also imagine this situation as follows. During the two components' "in-system" performance, component u_1 creates a situation in which component u_2 is "constantly bombarded" by a string of harmful "micro-shocks" (see Filus and Filus [5]). Each such micro-shock causes a corresponding "micro-damage" in the affected component u_2's physical constitution. We also assume that these micro-damages in component u_2's inner physical structure "cause" some corresponding "micro-changes in the original (baseline) failure rate" (and, in parallel, in the corresponding probability distribution) of its life-time X_2. After a, possibly long, time period X_1 of such interaction all these micro-damages cumulate their effects. As a result of this accumulation, the overall change in the corresponding "hazard rate function" will become significant. To describe formally the change in the hazard rate function we have chosen to consider corresponding changes in its parameter(s). In what follows we present a particular bivariate model for a 2-component system reliability which we called FF-Weibullian (formerly "pseudo-Weibullian" in Filus and Filus [4]).

Suppose the lifetimes of the components u_1 and u_2 in "laboratory conditions" are independent and distributed according to the Weibull density random variables X_1 and X_2.

Let $g_k(x_k) = \lambda_k \alpha_k x_k^{\alpha_k - 1} \exp{-\lambda_k x_k^{\alpha_k}}$ be the pdf of X_k (k = 1, 2).

Here, for k = 1, 2, we have the "vector parameter" $\theta_k = (\lambda_k, a_k)$.

Next consider the components u_1 and u_2 as acting within the system. Let the resulting (changed) values $\lambda_2^*; \sigma_2^*$ of the parameters of the (origi-

Parameter Dependence in Stochastic Modeling—Multivariate

nal) pdf $g_2(x_2; l_2, a_2)$ be determined by the following continuous functions of x_1:

$$\lambda_2^* = \lambda_2^*(x_1) \text{ and } \alpha_2^* = \alpha_2^*(x_1).$$

One then obtains the wide class of bivariate FF-Weibullian densities:

$$g(x_1,x_2) = g(x_1)g_2(x_2|x_1) = \lambda_1\alpha_1 x_1^{\alpha_1-1} \exp(-\lambda_1 x_1^{\alpha_1}) \lambda_2^*(x_1)\alpha_2^*(x_1) x_2^{\alpha_2^*(x_1)-1} \exp(-\lambda_2^*(x_1) x_2^{\alpha_2^*(x_1)})$$

(3)

where, for ease of computation, we recommend to apply as "sub-model" the following family of "parameter functions":

$$\lambda_2^*(x_1) = \lambda_2(1 + Ax_1^r) \text{ and } \alpha_2^*(x_1) = s\alpha_2 \text{ with parameters A, r, s positive reals.}$$

In particular, s may depend on x_1.

Another analytically interesting "sub-model" is given by:

$$\lambda_2^*(x_1) = \lambda_2 \exp[Ax_1^r] \text{ with A and r real, and } \alpha_2^*(x_1) = s\alpha_2(1 + cx_1) \text{ with } s > 0, c \geq 0.$$

Note that both factors $g_1(x_1)$ and $g_2(x_2|x_1)$ of the joint density $g(x_1, x_2)$ given by (3) are Weibullian densities. In particular, $g_2(x_2|x_1)$ is Weibullian with respect to the argument x_2 alone.

For the simpler FF-exponential example, see [1].

NOTICE ON SIMILAR INVESTIGATIONS IN THE LITERATURE

A parallel and basically independent path of investigation, which also has its roots in the bivariate normal distribution's dependence paradigm, is present in the literature under the key word "conditioning".

This method, used in the construction of numerous multivariate probability distributions, was extensively developed mostly since around 1987. See, for example, Arnold, Castillo and Sarabia [9] with citations. Also consider Castillo and Galambos [10] .

The underlying method (by numerous authors called the "conditioning method") relies on imposing conditional structure X|Y and Y|X on, given in advance, "baseline" probability densities f(x; A) and g(y; B) of some ("initially independent") random variables X and Y respectively, where A and B are scalar or vector parameters. The two conditional densities are defined as we did above, i.e.

g(y|x) = g(y; B(x)) and f(x|y) = f(x; A(y))

where A(y) and B(x) are continuous functions of realizations of the random variables Y, X respectively.

In this case the task is to find two proper (unknown) marginal densities for the bivariate probability distribution of (X, Y) which are, as a rule, not unique and sometimes do not exist.

Despite similarities this method essentially differs from ours. In our case, instead of the two conditional densities g(y|x) and f(x|y), we define only one, say, g(y|x), but together with the marginal f(x).

Pursuing this way we always directly obtain a unique model simply as the product of the two (known) densities.

In such a way, we have obtained a wide class of bivariate densities which is essentially disjoint from the class obtained by that alternative method. Also, the physical interpretation of the, so defined, conditional densities differs in the two approaches. However, both approaches are devoted to the same purpose which is to extend of the paradigm of the bivariate normal in stochastic modeling. Nevertheless, using the conditioning method it is very difficult to construct the multivariate distributions of any higher than two dimensions.

Parameter Dependence in Stochastic Modeling—Multivariate

Practically that method reduces to the bivariate cases while the method we present has a remarkable easiness of construction of probability distributions of, actually, arbitrary finite dimension. There is, namely, a recurrence procedure which allows to construct any j-th dimensional pdf based on corresponding (j – 1)-th dimensional pdf (j=3,4,...) constructed "at an earlier stage". That procedure was also used in Filus and Filus [11] for the construction of discrete time stochastic processes.

The next section is devoted to the construction of multivariate distributions for any arbitrary finite dimension.

METHOD OF PARAMETER DEPENDENCE FOR MULTIVARIATE PROBABILITY DISTRIBUTIONS CONSTRUCTION

For the construction, mentioned in the title, we successively use the simple recurrence method that yields the j-th dimensional probability density, given the (j – 1)-th one. Realize, that (for j = 3) we have already defined the 2- dimensional densities $g^2(x_1, x_2)$ by means of the products $g_1(x_1)g_2(x_2|x_1)$, where each underlying conditional density was given by $g_2(x_2|x_1) = g_2(x_2, q_2(x_1))$.

So the "first step" is already done. Suppose now that we have at our disposal the (j – 1)-th dimensional (j ≥ 3) pdf, say $g^{j-1}(x_1, x_2,...,x_{j-1})$ of the random vector $(X_1, X_2,...,X_{j-1})$. The task of obtaining the joint density $g_j(x_1, x_2,...,x_j)$ of the random vector $(X_1, X_2,...,X_{j-1}, X_j)$ always reduces to defining the conditional density $g_j(x_j|x_1,x_2,...,x_{j-1})$, given any univariate baseline pdf $h_j(x_j)$ by the method of parameter dependence. Assuming that originally q_j is a constant parameter, we define the conditional pdf $g_j(x_j|x_1,x_2,...,x_{j-1})$ by setting (according to the new physical situation of "being in the system" together with the other j – 1 objects):

$$g_j\left(x_j|x_1,x_2,\cdots,x_{j-1}\right) = h_j\left(x_j, \theta_j^*\left(x_1, x_2,\cdots,x_{j-1}\right)\right)$$

Now, the "new" value θ_j^* of the parameter is a continuous function of the realizations ("multi-stresses") $(x_1, x_2,...,x_{j-1})$ of the random vector $(X_1, X_2,...,X_{j-1})$.

The j-dimensional pdf of the random vector $(X_1, X_2,...,X_{j-1},X_j)$ one obtains simply as the product:

$$g^j(x_1,x_2,\cdots,x_j) = g_j(x_j|x_1,x_2,\cdots,x_{j-1})g^{j-1}(x_1,x_2,\cdots,x_{j-1})$$

The latter pdf becomes the basis for identical construction of the (j + 1)-dimensional pdf and so on.

We then stop the procedure once j + 1 = m, where m is the total dimension of the considered (maximal) random vector, say, $(X_1,...,X_m)$.

Since the analogy with the construction of each j-dimensional normal pdf (j=3,4,...) is not as straightforward as in the bivariate case, we found it beneficial to show this analogy closer, by reducing the normal's construction to the "diagonal case". Let us start with recalling that any normally distributed random vector, say, $X=(X_1,...,X_j)$, j=2,3,... ,m is obtainable from the random vector $Z=(Z_1,...,Z_j)$ by an affine transformation

$$X^T = AZ^T + \mu^T, \tag{4}$$

where the random variables $Z_1,...,Z_j$ are independent and each having the standard normal N(0, 1) pdf. A is an arbitrary j × j matrix with real entries (here, without losing generality, we restrict ourselves to nonsingular matrices A, only) and $\mu=(\mu_1,...,\mu_j)$ is an arbitrary fixed vector in R^j. The symbol ^{T} denotes the usual matrix transpose. Recall that every matrix A can be decomposed as

A=MB (5)

where B is a lower triangular and M is an orthogonal matrix. From (5) we obtain that any nonsingular lower triangular matrix B can be represented as the product:

Parameter Dependence in Stochastic Modeling—Multivariate

$$B = M^{\mathrm{T}} A, \tag{6}$$

where A is an arbitrary nonsingular matrix. If we replace representation (4) of the random normal vector X by the following representation

$$Y^{\mathrm{T}} = BZ^{\mathrm{T}} + \mu^{\mathrm{T}}, \tag{7}$$

then we replace the arbitrary random vector X by an arbitrary "triangular" random vector Y related to X by:

$$(Y - \mu)^{\mathrm{T}} = M^{\mathrm{T}} (X - \mu)^{\mathrm{T}}, \tag{8}$$

where M^{T} is an arbitrary orthogonal transformation.

Since the two zero-expectation random vectors $X - \mu, Y - \mu$ are obtained one from another by an isometry (here, rotation) M^{T} in the Euclidean space R^j, they may be considered as representing the same "stochastic data" expressed in two different (but still rectangular) coordinate systems. So from a stochastic viewpoint the "difference" between the random vectors X and Y is inessential and we can consider the random vector Y as an "arbitrary normal" ("with accuracy to the rotation" M^{T}).

Collecting all the above, we will consider the normal random vector Y, given by (7), where matrix B is any lower triangular matrix and Z is the standard normal j-vector. Write (7) in the form:

$$Y_1 = b_{11} Z_1 + \mu_1$$

$$Y_2 = b_{21} Z_1 + b_{22} Z_2 + \mu_2$$

$$\vdots$$

$$Y_{j-1} = b_{j-1,1} Z_1 + \cdots + b_{j-1, j-1} Z_{j-1} + \mu_{j-1}$$

$$Y_j = b_{j,1} Z_1 + \cdots + b_{j, j-1} Z_{j-1} + b_{j,j} Z_j + \mu_j \quad j \to m \tag{7*}$$

where m ³ j is the actual dimension of the constructed (final) random vector, say $(X_1,\ldots,X_m)$ (if m = ¥, one defines, in effect, a stochastic process with time j).

Considering the first -1 lines in (7*) as a system of linear equations, one obtains all $Z_1,\ldots,Z_{j-1}$ as linear combinations of $Y_1,\ldots,Y_{j-1}$. Substituting these solutions back into (7*) one obtains the following form:

$$Y_1 = c_{11}Z_1 + \mu_1$$

$$Y_2 = c_{22}Z_2 + (c_{21}Y_1 + \mu_2)$$

$$Y_3 = c_{33}Z_3 + (c_{31}Y_1 + c_{32}Y_2 + \mu_3)$$

$$\vdots$$

$$Y_{j-1} = c_{j-1,j-1}Z_{j-1} + (c_{j-1,1}Y_1 + \cdots + c_{j-1,j-2}Y_{j-2} + \mu_{j-1}) \qquad (9)$$

Realize that transformation (9) is easily reversible.

Assuming that realizations $y_1,\ldots,y_{k-1}$ of the random variables $y_1,\ldots,y_{k-1}$ are known, we obtain for each k=1,...,j:

$$Z_k = \frac{Y_k - (c_{k,1}y_1 + \cdots + c_{k,k-1}y_{k-1} + \mu_k)}{c_{kk}} \qquad (10)$$

where from the above assumed nonsingularity we have $c_{kk} \neq 0$. From (10) it follows that the conditional density of each Y_k, given the values $y_1,\ldots,y_{k-1}$, is normal and for the corresponding conditional expectation we have

$$E\left[Y_k \mid y_1,\ldots,y_{k-1}\right] = \mu_k + c_{k,1}y_1 + \cdots + c_{k,k-1}y_{k-1}$$

while for the (constant) conditional variance we obtain

$$\mathrm{Var}\left[Y_k \mid y_1,\ldots,y_{k-1}\right] = c_{kk}^2$$

Parameter Dependence in Stochastic Modeling—Multivariate

To adopt the above procedure to our concept of "baseline" T_j versus "in system" Y_j random variables, replace in (9) the independent standard random variables $Z_1, Z_2, \ldots$ by independent random variables, say, $T_1, T_2, \ldots$, where each T_k has the ("baseline") normal $N(\mu_k, s_k)$ pdf ($k=1,2,\ldots$).

Replace transformation (7) by

$$Y = BT^T, \qquad (11)$$

Where $T=(T_1,\ldots,T_j)$. Using this change, (10) will be replaced by the following inverse transformation:

$$T_k = \frac{Y_k - (c_{k,1} y_1 + \cdots + c_{k,k-1} y_{k-1})}{c_{kk}} \qquad (10^*)$$

This yields the conditional pdf of $Y_k | y_1, \ldots, y_{k-1}$ to be the normal

$$N\left(\mu_k + c_{k,1} y_1 + \cdots + c_{k,k-1} y_{k-1}, |c_{kk}| \sigma_k \right)$$

Finally, the general pattern of "creation" of any successive j-variate normal pdf can be explained as follows.

Given are the first $j - 1$ lines of transformation (11) in the form:

$$Y_1 = c_{11} T_1$$
$$Y_2 = c_{22} T_2 + c_{21} Y_1$$
$$Y_3 = c_{33} T_3 + (c_{31} Y_1 + c_{32} Y_2)$$
$$\vdots$$
$$Y_{j-1} = c_{j-1, j-1} T_{j-1} + (c_{j-1,1} Y_1 + \cdots + c_{j-1, j-2} Y_{j-2}) \qquad (12)$$

for some $j-1=1,2$. (Realize that the joint normal pdf $g_{j-1}(y_1, y_2, \ldots, y_{j-1})$ of the random vector $(Y_1, Y_2, \ldots, Y_{j-1})$ was defined in $j - 1$ "previous" steps.

In particular for $j-1=1$, it is the univariate normal $N(\mu_1, |c_{11}|s_1)$ density of the variable Y_1.)

We may assume that the next baseline random variable T_j, originally having the $N(\mu_j, s_j)$ pdf, is incorporated to the "system" by transforming

$$T_j \to Y_j | Y_1, \cdots, Y_{j-1}$$

This transformation is thought of as adding to (12) the following j-th line:

$$Y_j = c_{j,j} T_j + \left(c_{j,1} Y_1 + \cdots + c_{j,j-1} Y_{j-1} \right) \tag{13}$$

["Physically" this could mean that the variables $Y_1, \ldots, Y_{j-1}$ "become" explanatory ("stresses") for the "new" variable Y_j obtained from T_j that (originally) was independent from these stresses].

From (13) one can determine the conditional pdf of Y_j, given any realization $(y_1, \ldots, y_{j-1})$ of the random vector $(Y_1, \ldots, Y_{j-1})$, as the following normal pdf in y_j:

$$N\left(\mu_j + c_{j,1} y_1 + \cdots + c_{j,j-1} y_{j-1}, |c_{jj}| \sigma_j \right).$$

Thus, as the j-th "object" (originally independent from the "system" and characterized by the random quantity T_j) was "put into the system" the quantity T_j turns to the quantity Y_j and, in parallel, the parameters μ_j and s_j of its normal density are turned into $\mu_j + c_{j,1} y_1 + \ldots + c_{j,j-1} y_{j-1}$ and $\sigma_j |c_{jj}|$, respectively, while normality is preserved.

Clearly, the new value μ_j^* of the (conditional) expectation became the continuous (here linear) function

$$\mu_j^*\left(y_1, \cdots, y_{j-1} \right) = \mu_j + c_{j,1} y_1 + \cdots + c_{j,j-1} y_{j-1}$$

Parameter Dependence in Stochastic Modeling—Multivariate

of realizations $y_1, y_2, \ldots, y_{j-1}$ while unfortunately the new value of the standard deviation does not depend on $y_1, y_2, \ldots, y_{j-1}$ but remains constant even if multiplied by the specific, determined by the "system", number c_{jj}. This can be "made up" if we allow the number c_{jj} in (13) to be dependent on $y_1, y_2, \ldots, y_{j-1}$, but then the so obtained multivariate FF-normal distribution ceases to be normal since (13) ceases to be linear.

As we have shown also in multivariate cases, the origin of the "parameter dependence method for the construction", lies in the construction of the multivariate normal distributions. Recall that having defined the conditional pdf $g_j(y_j | y_1, \ldots, y_{j-1})$ and the joint pdf $g^{j-1}(y_1, \ldots, y_{j-1})$ we automatically have the joint pdf $g^j(y_1, \ldots, y_j)$ as the simple arithmetic product of the two. In the case just considered, all the densities $g^j(y_1, \ldots, y_j), j=1, 2, \ldots,$ are (arbitrary with the accuracy to the rotations in R^j) multivariate normal.

Preserving the general spirit of the multivariate normal pdf derivation, let us extend all the Equations (13) for $j=2,3,\ldots,m$ by allowing the translations $c_{j,1} Y_1 + \ldots + c_{j,j-1} Y_{j-1}$ to be any nonlinear continuous function of $Y_1, \ldots, Y_{j-1}$ and replacing the constant c_{jj} by any continuous function of the same variables. Now, for any $j=2,3,\ldots,m$ (13) may be rewritten into the following "triangular" (see Filus, Filus and Arnold [12]) form:

$$Y_j = \Phi_j(Y_1, \cdots, Y_{j-1}) T_j + \Psi_j(Y_1, \cdots, Y_{j-1}) \qquad (13^*)$$

where $\Phi_j()$ and $\Psi_j()$ are arbitrary continuous functions and $P(\Phi_j(Y_1, \cdots, Y_{j-1}) \neq 0) = 1$.

From (13*) we obtain its inverse:

$$T_j = \frac{Y_j - \Psi_j(Y_1, \cdots, Y_{j-1})}{\Phi_j(Y_1, \cdots, Y_{j-1})} \qquad (13^{**})$$

and then for each observation $(y_1, \ldots, y_{j-1})$ of $(Y_1, \ldots, Y_{j-1})$ the conditional pdf of $Y_j | y_1, \ldots, y_{j-1}$ as follows:

$$g_j(y_j|y_1,\ldots,y_{j-1}) = N\left(\mu_j + \Psi_j(y_1,\ldots,y_{j-1}); \sigma_j|\Phi_j(y_1,\ldots,y_{j-1})|\right) \quad (14)$$

It is clear that the sequence of the densities (14) (j=2,3,…,m) together with the normal initial density $g_1(y_1)$ of Y_1 uniquely determines the m-variate FF-normal pdf $g(y_1,\ldots,y_m)$ of the random vector $(Y_1,\ldots,Y_m)$. Remarkably, this non-normal density has its natural representation as the product of m normal densities:

$$g(y_1,\ldots,y_m) = g_1(y_1)\prod_{j=2}^{m} g_j(y_j|y_1,\ldots,y_{j-1}) \quad (15)$$

However, the marginal pdfs of $Y_2,\ldots,Y_m$ are not normal anymore.

The main conclusion which follows the considerations in Sections 6.2 - 6.4 may be stated as: There is a generic relationship that associates the construction method of the parameter dependence with the stochastic dependence structure present within the multivariate normal distribution of any dimension.

As an example of this relationship realize that the transformations (13) and (13*), when applied to the independent normal random variables T_j (j=1,2,…,m), define multivariate normal and FF-normal pdfs respectively. They produce other m-variate probability distributions if the normality assumption for T_j is dropped.

Let now T_j (j=1,2,…,m) be independent random variables all having the standard exponential pdf:

$$f_j(t_j) = \exp[-t_j].$$

Applying to the random vector $(T_1,\ldots,T_m)$ transformation (13*) and then (13**) for j=1,2,…,m, one obtains the joint density $g(y_1,\ldots,y_m)$ of the resulting random vector $(y_1,\ldots,y_m)$ to be given by the product of m factors (15), where according to first row of (12) we have

$$g_1(y_1) = (1/|c_{11}|)\exp[-y_1/|c_{11}|]$$

Parameter Dependence in Stochastic Modeling—Multivariate

and according to (13**)

$$g_j(y_j|y_1,\cdots,y_{j-1}) = \frac{1}{|\Phi_j(y_1,\cdots,y_{j-1})|} \exp\frac{-(y_j - \Psi_j(y_1,\cdots,y_{j-1}))}{|\Phi_j(y_1,\cdots,y_{j-1})|}$$

where the latter is the two parameter exponential density with respect to y_j for j=2,3,...,m.

Another interesting case of the m-variate FF-Weibullian pdf can be obtained by applying transformations (13*) to m independent Weibullian random variables. An even more general class of FF-Weibullians one obtains using the pseudopower transformations (see Filus and Filus [4]) instead of the pseudoaffine (13*) which actually is a special case of the pseudopower.

All these distributions (including the m-variate normal) can as well be obtained by direct use of the "parameter dependence pattern" which produces more m-variate models than the considered above transformations. On the other hand existence of the defining transformations facilitates an underlying statistical analysis and simulations.

OTHER PARAMETER DEPENDENCE PARADIGMS IN THE LITERATURE

Some paradigms, applied in the reliability literature, are exactly those of the "parameter dependence" that we describe in this paper. However, in most of the cases they are not directly related to the problem of construction of multivariate probability distributions (so, also are different from the "conditioning" procedures in [9] [10] ; see above, Section 5). There are two such subjects that we discuss in the following.

The Accelerated Life Testing

When testing the life times of some high reliability products, the stresses usually encountered such as temperature, humidity, voltage sometimes are kept on significantly higher than usual levels in order to make the life times shorter than they are in normal conditions. The so obtained data (a "sample") is then extrapolated into those (hypothetical) life times that would, possibly, be obtained under the regular values of the stresses. Existence of rules, that associate the products' life times with values of the stresses applied, is necessary for performing proper extrapolations. Several such rules, typically known as the Arrhenius or Eyring (see, Meeker and Escobar [13] , Nelson [14] and [15] an internet source) relationships, are based on physical and chemical considerations on the rates of some chemical reactions that give rise to a given unit's failure. The obtained models, in general, allow determining the ratio of life times of the same product under higher and under normal temperature or, in the case of the Eyring model, some other physical quantities that play the role of the stresses (see, [13] , formula (18.5) page 476). Methods like that (i.e., the so called SAFT models [13]) directly relate the (life) times by means of a simple coefficient called the "acceleration factor".

Unfortunately, with this method the simplicity often comes along with inaccuracy of the predictions. Other methods apply the "Proportional Hazards Relationships" known also as Cox Model (see Cox [16]) which instead of times relate hazard rates.

More recently ([15]), the relationships between the life times under different stresses are related indirectly through their probability distributions via distribution's parameter in that way that considers distribution's parameter as a function of a given stress (mostly temperature, humidity, pressure, voltage). Those relationships, even if considered in a different context, obey the same paradigm (of the parameter dependence) as that considered in this paper in association with the construction of the multivariate probability distributions.

In what follows we discuss the differences.

1) The generality of the "parameter dependence theory" we built in this paper, is significantly higher than the very special case applied to the accelerated life testing theory. There are three reasons for that.

Firstly, in our approach the subject of constructing conditional probability distributions is not limited to the life testing, and not even to the "stress-life time" pattern only. The range of applications of our theory is very wide, including many biomedical (see, Collett [17]) and econometric relationships, (see, Filus and Filus [11] also Filus, Filus and Krysiak [18]).

Secondly, in the paradigm we consider, the relation between a parameter and stress (or any other random quantity) is given by an arbitrary continuous function, while the number of such functions applied in association with the accelerated life time testing is very limited. Actually, the functions are restricted to few "models" such as the Arrhenius, Eyrie, inverse power law, log-linear, and not many more (see, for example, the Eyring-Weibull model in [15] , Section 5). Those models were obtained from physical and chemical considerations that are only valid for some simple failure (degradation) mechanisms while very often real failure mechanism is too complicated to be analyzed that way.

Our idea is to omit the complicated physical or chemical phenomena that often are poorly understood and to apply two steps purely empirical approach.

Speaking roughly, the first step is an "educated guess" (for choice of a proper function) and the second is statistical verification of this guess.

Thirdly, in our theory we may consider an arbitrary parameter of an arbitrary probability distribution as a stress dependent, while, according to our knowledge (see, for example, [15]), the only life time distributions so far considered in the accelerated life testing are exponential, Weibullian, and lognormal (normal), and for each the distribution only one parameter is taken under consideration.

2) Besides the generality (of the constructed conditional distributions) our concept also differs with regard to the purpose. Namely, independently of the conditional distributions construction, we also have the construction of bivariate and multivariate probability distributions such as the FF-normal, FF-exponential, FF-Weibullian, FF-gamma and other

(for comparison with similar "conditioning methods" of construction present in the literature, see Section 5).

The construction of high dimension multivariate distributions based on parameter dependence can easily be extended to Markovian and non-Markovian (still simple!) stochastic processes (see Filus and Filus [11]). The latter constructions seem to be rather unique in the literature.

Load Control and Load Sharing

1. Other than the accelerated life testing subject, where the "parameter dependence paradigm" is applied, is a set of problems centered around the notion of "load optimization" (see Filus [19] [20] , Levitin and Amari [21] , Nourelfath and Yalaoui [22] and others). This relatively new topic can be described as follows. Some working systems, such as cargo transporting trucks, trains, electric power lines, highways, computer processors, or other systems supporting varying amounts of load, require control of that load. On the one hand more load yields more gain, but, on the other hand, load (as stress) increment yields a corresponding increment of the system's failure rate which, in turn, depends on certain parameters [20] . The optimization problem can be formulated in several ways ([20] [21]) but, in all cases, the main idea is to balance between an expected gain that follows a good (load) transportation and the loss in reliability of the transporting medium which decreases the overall gain. Thus, for some (optimal) value of the load to be found, the pure expected gain earned by the system is maximal. In this framework the relationship between stress (the load) and some parameters of the system life time probability distribution (failure rate) is vital for finding a proper model. As examples of such relations, the power and exponential functions of the load were chosen in [20] .
2. Similar application of the parameter dependence pattern also occurs when "load sharing phenomena" takes place. Suppose that we have a parallel system supporting a load such as several engines aircraft or two electric power lines. Failure of any system's component may cause the total load to be redistributed among fewer components, so that the load on each of them increases by some predictable value. Now we may encounter either the load optimization problem [21] or simply the task of determination of system's life time probability distribution (see Freund [23] for the exponential case as well as Lu [24] and Filus [25] for Weibullian and lognormal cases). In all these cases the parameter dependence pattern is involved or it is desirable to apply it to get a deeper insight into the underlying stochastic phenomena.

Remark. As a final remark, let me mention the relationship between the parameter dependence presented in this paper, and the stochastic dependence based on models initiated in 1961 by Freund [23] . Besides some similarity the most basic difference lies in the fact that in the Freund scheme the system components act independently until the first failure. In general, the components successive failures cause the total load to be shared by fewer and fewer remaining components, affecting their failure rates (via the parameters).

Quite opposite to that, in the models we introduce, the component interactions take place only when the components work. Any failure of a system component stops its influence on the remaining components' life times. Therefore, the two paradigms, the "Freund's load sharing" and our "parameter dependence", are "disjoint" and in a sense "complementary". In reality, both (physical) phenomena may take place at the same time and it seems to be quite possible in the future to construct stochastic models (i.e., multivariate probability distributions) that would obey both paradigms.

Nevertheless, we stress the generic relation of all the multivariate probability distributions based on the parameter dependence with the multivariate Gaussians.

REFERENCES

1. Filus, J.K. and Filus, L.Z. (2013) A Method for Multivariate Probability Distributions Construction via Parameter Dependence. Communications in Statistics: Theory and Methods, 42, 716-721. http://dx.doi.org/10.1080/03610926.2012.731549
2. Filus, J.K. and Filus, L.Z. (2000) A Class of Generalized Multivariate Normal Densities. Pakistan Journal of Statistics, 16, 11-32.
3. Filus, J.K. and Filus, L.Z. (2007) On New Multivariate Probability Distributions and Stochastic Processes with Systems Reliability and Maintenance Applications. Methodology and Computing in Applied Probability, 9, 426-446,
4. Filus, J.K. and Filus, L.Z. (2006) On Some New Classes of Multivariate Probability Distributions. Pakistan Journal of Statistics, 1, 21-42.

5. Filus, J.K. and Filus, L.Z. (2008) On Multicomponent System Reliability with Microshocks-Microdamages Type of Components' Interaction. Proceedings of the International Multiconference of Engineers and Computer Scientists, Lecture Notes in Engineering and Computer Science, Hong Kong, 19-21 March 2008, 1945-1951.
6. Filus, J.K. and Filus, L.Z. (2010) Weak Stochastic Dependence in Biomedical Applications. American Institute of Physics Conference Proceedings 1281, Numerical Analysis and Applied Mathematics, III, 1873-1876.
7. Kotz, S., Balakrishnan, N. and Johnson, N.L. (2000) Continuous Multivariate Distributions (Vol. 1). 2nd Edition, J. Wiley & Sons, Inc, New York, 217-218.http://dx.doi.org/10.1002/0471722065
8. Barlow, R.E. and Proschan, F. (1975) Statistical Theory of Reliability and Life Testing. Holt, Rinehart and Winston, New York.
9. Arnold, B.C., Castillo, E. and Sarabia, J.M. (1999) Conditional Specification of Statistical Models. Springer Series in Statistics, Springer Verlag, New York.
10. Castillo, E. and Galambos, J. (1990) Bivariate Distributions with Weibull Conditionals. Analysis Mathematica, 16, 3-9. http://dx.doi.org/10.1007/BF01906769
11. Filus, J.K., Filus, L.Z. (2008) Construction of New Continuous Stochastic Processes. Pakistan Journal of Statistics, 24, 227-251.
12. Filus, J.K., Filus, L.Z. and Arnold, B.C. (2010) Families of Multivariate Distributions Involving "Triangular" Transformations. Communications in Statistics—Theory and Methods, 39, 107-116.
13. Meeker, W.Q. and Escobar, L.A. (1998) Statistical Methods for Reliability Data. John Wiley & Sons, Inc., New York.
14. Nelson, W. (1990) Accelerated Testing: Statistical Models, Test Plans and Data Analysis. Wiley, New York. http://dx.doi.org/10.1002/9780470316795
15. (2013) www.ReliaWiki.org/index.php/Accelerated_Life_Testing_Data_Analysis_Reference
16. Cox, D.R. (1972) Regression Models and Life Tables (with Discussion). Journal of the Royal Statistical Society, B74, 187-220.
17. Collett, D. (2003) Modeling Survival Data in Medical Research. 2nd Edition, Chapman @ Hall/CRS A CRC Press Company, London, New York, Washington, D.C.
18. Filus, J.K., Filus, L.Z. and Krysiak, Z. (2013) Analytical Statistical and Simulation Models Utilized in Modeling the Risk in Finance. 5th International Conference on Risk Analysis, Tomar, Portugal, 30th May-1st June.
19. Filus, J.K. (1986) A Problem in Reliability Optimization. Journal of the Operational Research Society, 37, 407-412.
20. Filus, J.K. (1987) The Load Optimization of a Repairable System with Gamma Distributed Time-to-Failure. Reliability Engineering, 18, 275-284. http://dx.doi.org/10.1016/0143-8174(87)90032-1

21. Levitin, G. and Amari, S.V. (2009) Optimal Load Distribution in Series-Parallel Systems. Reliability Engineering and System Safety, 94, 254-260. http://dx.doi.org/10.1016/j.ress.2008.03.001
22. Nourelfath, M. and Yalaoui, F. (2012) Integrated Load Distribution and Production Planning in Series-Parallel Multistate Systems. Reliability Engineering and System Safety, 106, 138-145. http://dx.doi.org/10.1016/j.ress.2012.06.006
23. Freund, J.E. (1961) A Bivariate Extension of the Exponential Distribution. Journal of the American Statistical Association, 56, 971-977. http://dx.doi.org/10.1080/01621459.1961.10482138
24. Lu, J. (1989) Weibull Extensions of the Freund and Marshal-Olkins Bivariate Exponential Models. IEEE Transactions on Reliability, 38, 615-619. http://dx.doi.org/10.1109/24.46492
25. Filus, J.K. (1991) On a Type of Dependencies between Weibull Life times of System Components. Reliability Engineering and System Safety, 31s, 267-280. http://dx.doi.org/10.1016/0951-8320(91)90071-E

CITATION

Filus, J. and Filus, L. (2014) Parameter Dependence in Stochastic Modeling—Multivariate Distributions. *Applied Mathematics*, 5, 928-940. doi: 10.4236/am.2014.56088.

Set-Valued Stochastic Integrals with Respect to Finite Variation Processes

Jinping Zhang and Jiajia Qi
Department of Mathematics and Physics, North China Electric Power University, Beijing, China

ABSTRACT

In a Euclidean space R^d, the Lebesgue-Stieltjes integral of set-valued stochastic processes $F = \{F_t(\omega), t \in [0,T]\}$ with respect to real valued finite variation process $\{A_t(\omega), t \in [0,T]\}$ is defined directly by employing all integrably bounded selections instead of taking the decomposable closure appearing in some existed references. We shall show that this kind of integral is measurable, continuous in t under the Hausdorff metric and L^2-bounded.

INTRODUCTION

Recently, integrals for set-valued stochastic processes with respect to Brownian motion, martingales and the Lebesgue measure have received much attention.

In 1997, Kisielewicz ([1]) defined the integral of setvalued process as a subset of L^2 space, but he didn't consider the measurability of the integral. In 1999, Kim and Kim [2] used the definition of stochastic integrals of set-valued stochastic process with respect to the Brownian motion. They called it Aumann ([3]) type Itô integrals. In [4], Jung and Kim modified the definition by taking the decomposable closure such that the integral is measurable. Li and Ren [5] modified Jung and Kim's definition by considering the predictable set-valued stochastic process as a set-valued random variable in the product space $(R_+ \times \Omega)$, and the measurability and decomposability also were based on product σ-algebra. After that, Zhang et al. ([6, 7]) studied the set-valued integrals with respect to the martingale and Brownian motion.

Stochastic differential inclusions and set-valued stochastic differential (or integral) equations are employed to model the problems with not only randomness but also impreciseness. Recently, there are some references related to set-valued differential equations such as [8-13] etc.

Concerning to the integral with respect to finite variation processes, Malinowski and Michta [12] give the notion of set-valued integral with respect to single valued finite variation but without considering the measurability. Z.Wang and R.Wang [14] defined the Lebesgue-Stieltjes stochastic integral of single valued stochastic processes with respect to set-valued finite variation processes (refer to [14] for the detail).

In this paper, different from the definition in [14], based on the Definition 3.1 in [12], we will study the Lebesgue-Stieltjes integral of set-valued stochastic processes with respect to single valued finite variation process. We shall prove the measurability of integral, namely, it is a set-valued random, which is similar to the classical stochastic integral.

This paper is organized as follows: in section 2, we present some notions and facts on set-valued random variables; in section 3, we shall give the definition of integral of set-valued stochastic processes with respect to finite variation process and then prove the measurability and L^2-boundedness.

PRELIMINARIES

We denote N the set of all natural numbers, R the set of all real numbers, R^d the d-dimensional Euclidean space with the usual norm $\|\cdot\|$, R_+ the set of all nonnegative numbers. Let (Ω, F, P) be a complete prob-

Set-Valued Stochastic Integrals with Respect to Finite Variation

ability space, $\{F_t : t \in [0,T]\}$ a σ-field filtration satisfying the usual conditions. Let B(E) be a Borel field of a topological space E.

Let $K(R^d)$ (resp. $K_k(R^d), K_{kc}(R^d)$) be the family of all nonempty, closed (resp. nonempty compact, nonempty compact convex) subsets of R^d. For any $x \in R^d$ and $A \in K(R^d)$, define the distance between x and A by $d(x,A) = \inf_{y \in A} \|x-y\|$. The Hausdorff metric d_H on $K(R^d)$ (see e.g. [15]) is defined by

$$d_H(A,B) = \max\left\{\sup_{a \in A} d(a,B), \sup_{b \in B} d(b,A)\right\} \qquad (1)$$

$A, B \in K(R^d)$.

Denote $\|A\|_k = d_H(\{0\}, A) = \sup_{\alpha \in A}\|\alpha\|$. For $A, B, C, D \in K(R^d)$, we have

$$H(A+B, C+D) \leq H(A,C) + H(B,D).$$

For $A \subset R^d, x^* R^d$ the support function of A is defined as follows:

$$S(x^*, A) = \sup\{\langle x^*, x \rangle : x \in A\}.$$

$L^p(\Omega, F, P; R^d) = L^p(\Omega; R^d)(p \geq 1)$: the set of all R^d—valued Borel measurable functions $f : \Omega \to R^d$ such that the norm

$$\|f\|_p = \left\{\int_\Omega \|f(\omega)\|^p \, dP\right\}^{\frac{1}{p}}, \text{ if } 1 \leq p < \infty,$$

$$\|f\|_\infty = \text{ess}\sup_{\omega \in \Omega}\|f(\omega)\|, \text{ if } p = \infty,$$

is finite. F is called L^p-integrable if $f \in L^p(\Omega; R^d)$.

A set-valued function $F: \Omega \to K(R^d)$ is said to be measurable if for any open set $0 \subset R^d$, the inverse $F^{-1}(0) := \{\omega \in \Omega : F(\omega) \cap 0 \neq \phi$ belongs to F. Such a function F is called a set-valued random variable.

Let $M(\Omega, F, P; K(R^d))$ (resp. $M(\Omega, F, P; K_c(R^d))$, $M(\Omega, F, P; K_{Kc}(R^d))$) be the family of all measurable $K(R^d)$-valued (resp. $K_c(R^d)$, $K_{kc}(R^d)$-valued) functions, briefly by $M(\Omega; K(R^d))$ (resp. $M(\Omega, F, P; K_c(R^d))$, $M(\Omega, F, P; K_{Kc}(R^d))$). For $F \in M(\Omega, K(R^d))$, the family of all L^p-integrable selections is defined by

$$S_F^p(F) := \{f \in L^p(\Omega, F, P; R^d) : f(\omega) \in F(\omega) \text{ a.s.}\},$$
$$p \geq 1 \tag{2}$$

In the following, $S_F^p(F)$ is denoted briefly by S_F^p.

A set-valued random variable F is said to be integrable if S_F^1 is nonempty. F is called $L^p(p \geq 1)$-integrably bounded if there exits $h \in L^p(\Omega, F, P; R^d)$ s.t. for all $x \in F(\omega), \|x\| \leq h(\omega)$ almost surely.

An R^d-valued stochastic process $f = \{f_t : t \geq 0\}$ (or denoted by $f = \{f(t) : t \geq 0\}$) is defined as a function $f : R_+ \times \Omega \to R^d$ with the F-measurable section f_t, for $t \geq 0$. We say f is measurable if f is $B_+ \otimes F$ - measurable. The process $f = \{f_t : t \geq 0\}$ is called F_t-adapted if f_t is F_t-measurable for every $t \geq 0$. Let $\sum := I_{t \geq 0}\{Z \in B(R_+) \otimes F : Z_t \in F_t\}$, where $Z_t = \{\omega; (t, \omega) \in Z\}$. We know that $\sum$ is a σ-algebra on $R_+ \times \Omega$. A function $f : R_+ \times \Omega \to R^d$ is measurable and F_t-adapted if and only if it is $\sum$-measurable ([8]).

Set-Valued Stochastic Integrals with Respect to Finite Variation

In a fashion similar to the R^d-valued stochastic processes, a set-valued stochastic process $F = \{F_t : t \geq 0\}$ is defined as a set-valued function $F : R_+ \times \Omega \to K(R^d)$ with F-measurable section F_t for $t \geq 0$. It is called measurable if it is $B_+ \otimes F$-measurable, and F_t- adapted if for any fixed t, $F_t(\cdot)$ is F_t-measurable. $F = \{F_t : t \geq 0\}$ is measurable and F_t-adapted if and only if it is $\sum$-measurable. $F = \{F_t : t \geq 0\}$ is called L^p-integrable if every F^t is L^p-integrable.

SET-VALUED STOCHASTIC INTEGRAL W.R.T FINITE VARIATION PROCESSES

Let $A = \{A_t, t \geq 0\}$ be a real valued F_t-adapted measurable process with finite variation and continuous sample trajectories a.s. from the origin. That is to say, for each compact interval $[s,t] \subset [0,\infty)$ and any partition $\Delta = \{t_1, L, t_n\}$ of $[s,t]$, the total variation

$$V_A([s,t]) = \sup_\Delta \Sigma_{2=1}^n \left| A_{t_i}(\omega) - A_{t_{i-1}}(\omega) \right|$$

is finite and $A(0,\cdot) = 0$ a.s. Then for any T>0, the process $A = \{A_t, t \geq 0\}$ can generate a random measure denoted by μ_A in the space $[0,T], (B[0,T]))$. For any $(s,t] \subset [0,T]$, let

$$\mu_A((s,t]) := |A(t,\omega)| - |A(s,\omega)|$$

where $A(t,\omega) = A^+(t,\omega) + A^-(t,\omega)$ is the decomposition of A, A^+ and A^- are non-negative and nondecreasing processes, $|A(t,\omega)| = A^+(t,\omega) + A^-(t,\omega)$. In the product space $(\Omega \times [0,T], \sum)$, set

$$v_A(C) := \int_\Omega \int_{[0,T]} 1_C(t,\omega) \mu_A([0,T]) \mu_A(dt) P(d\omega), \tag{3}$$

For $C \in \Sigma$, where 1_c is the index function. Then the set function v is a finite measure in the measurable space $(\Omega \times [0,T], \Sigma)$ if and only if $\int_\Omega (\mu_A([0,T]))^2 P(d\omega) < \infty$. In the following we always assume $\int_\Omega (\mu_A([0,T]))^2 P(d\omega) < \infty$.

Let $L^2(\Omega \times [0,T], \Sigma, V_A; R^d)$ be the family of all Σ-measurable R^d-valued stochastic processes f such that

$$\int_{\Omega \times [0,T]} \|f(\omega,t)\|^2 v_A(dt) < \infty.$$

For any $f \in L^2(\Omega \times [0,T], \Sigma, V_A; R^d)$ and $[s,t] \subset [0,T]$, the stochastic Lebesgue-Stieltjes integral $\int_{[s,t]} f(\tau) dA_\tau$ is defined by the Bochner integral $\int_{[s,t]} f(\tau) \mu_A(d\tau)$ pathby-path. One can show that the integral process

$$\left\{ \int_{[0,t]} f(s) dA_s, t \in [0,T] \right\} \text{ is } \Sigma \text{ -measurable.}$$

Note: in [12], the integrand is assumed being predictable, in fact the integrand can be relaxed to the Σ - measurable class since the integrator A_t is continuous.

Let $M^2(\Omega \times [0,T], \Sigma, v_A; K(R^d))$ be the family of all Σ-measurable $K(R^d)$-valued stochastic processes F such that

$$\int_{\Omega \times [0,T]} \|F(\omega,t)\|_k^2 v_A(dt) < \infty,$$

Where

$$\|F(\omega,t)\|_k = \sup_{x \in F(\omega,t)} \|x\|$$

Set-Valued Stochastic Integrals with Respect to Finite Variation

For any

$$F \in M^2\left(\Omega \times [0,T], \Sigma, v_A; \mathcal{K}(R^d)\right)$$

Set

$$S^2(F) := \left\{ f \in M^2\left(\Omega \times [0,T], \Sigma, v_A; R^d\right) : f(\omega,t) \in F(\omega,t), v_A - a.e. \right\} \quad (4)$$

Definition 1: (see [12]) For a set-valued stochastic process $F \in M^2\left(\Omega \times [0,T]; K_{kc}(R^d)\right)$ the set-valued stochastic Lebesgue-Stieltjes integral (over interval [s,t]) of F with respect to the finite variation continuous process A is the set

$$\int_{[s,t]} F(\tau) dA_\tau := \left\{ \int_{[s,t]} f(\tau) dA_\tau : f \in S^2(F) \right\}.$$

In [12], the authors call this kind of integral as trajectory integral since they consider it as a

$K\left(L^2\left(\Omega \times [0,T], \Sigma, v_A; R^d\right)\right)$-valued random variable. Here, we shall consider it as a subset of R^d and show the measurability with respect to F, which is very different from the way in [12], also different from other references such as [10,16,17] etc. In fact, for almost every $\omega \in \Omega$, the above integral $\int_{[s,t]} F(\tau) dA_\tau$ is a subset of R^d. In the following, we shall assume the σ-algebra F is separable w.r.t P. In addition, B([0,T]) is separable and $\Sigma \subset F \otimes B([0,T])$, then one can get $S^2(F)$ is separable. Therefore we can find an F-measurable set Ω_F, such that $P(\Omega_F) = 1$ and for every

$\omega \in \Omega_F$, the integral $\int_{[0,T]} F(\tau)dA_\tau$ is defined path-bypath. For $\omega \in \Omega/\Omega_F$, set $\int_{[s,t]} F(\tau)dA_\tau = \{0\}$, therefore it is well defined for every $\omega \in \Omega$.

$\int_{[s,t]} F(\tau)dA_\tau = \int_{[s,t]} F(\tau)dA_\tau$ since the continuity of A_t. In the sequel, we shall denote the integral by $\int_s^t F(\tau)dA_\tau$ instead of $\int_{[s,t]} F(\tau)dA_\tau$. For any $t \in [0,T]$ denote $\int_0^t F_s dA_s$ by $\Gamma_t(F)$.

Theorem 1: For $F \in M^2\left(\Omega \times [0,T], \Sigma, K_{kc}(R^d); v_A\right); [s,t] \subset [0,T]$ and $\omega \in \Omega$, the Lebesgue-Stieltjes integral $\int_s^t F_\tau(\omega)dA_\tau(\omega)$ is a compact and convex subset of R^d.

Proof 1: In fact, $S^2(F)$ is a bounded and convex subset of $L^2\left(\Omega \times [0,T], \Sigma, R^d; v_A\right)$, since F is convex and compact, moreover, it is weakly compact since $L^2\left(\Omega \times [0,T], \Sigma, R^d; v_A\right)$ is reflexive. The convexity of the integral is obvious.

We shall show the linear operator $T(F) := \int_s^t F_\tau(\omega)dA_\tau(\omega) : S^2(F) \to K_{kc}(R^d)$ is bounded.

For any $f \in S^2(F)$, $[s,t] \subset [0,T]$,

$$\left\| \int_s^t f(\tau,\omega)dA_\tau(\omega) \right\| \leq \int_s^t \|f(\tau,\omega)\| dA_\tau(\omega)$$

$$\leq \int_s^t \|F(\tau,\omega)\|_K |dA_\tau(\omega)| < \infty, \tag{5}$$

Set-Valued Stochastic Integrals with Respect to Finite Variation

which implies the linear operator T is bounded. Therefore the integral $\int_s^t F_\tau dA_\tau$ is weakly compact since the bounded linear operator mapping a weakly compact set to a weakly compact one. In R^d space, a weakly compact set is compact.

Lemma 1: (see [16] Corollary 2.1.1 (5)) Assume (Ω, A) is a measurable space, X is a separable Banach space, $F: \Omega \to K(X)$, and F is a set-valued random variable, then $S(x^*, F(\omega))(x^* \in x^*)$ is measurable.

By using Lemma 1, as a manner similar to Theorem 1 in [17], we have the following result:

Lemma 2: Assume A is the corresponding stochastic process, $F \in M^2\left(\Omega \times [0,T], \Sigma, K_{kc}(R^d); v_A\right)$ for any $[s,t] \subset [0,T]$, we have

1. $\int_s^t \alpha F_\tau dA_\tau = \alpha \int_s^t F_\tau dA_\tau, \ \alpha \in R$;

2. $S\left(x^*, \int_s^t F_\tau dA_\tau\right) = \int_s^t S\left(x^*, F_\tau\right) dA_\tau, \ x^* \in R^d$.

Lemma 3: (see [16] Theorem 2.1.16) Assume (Ω, F) is a measurable space, X is a separable Banach space, $F: \Omega \to K_{kc}(X)$, and for any fixed $x^* \in X^*, S(x^*, F)$ is measurable, if one of the following conditions is satisfied:

1. X^* is separable;
2. for any $\omega \in \Omega, F(\omega) \in K_{kc}(X)$.

Then F is a set-valued random variable.

From Lemma 1 and Lemma 3, when $X = R^d$, for any $x^* \in R^d; F(\omega) \in K_{kc}(R^d)$ is F-measurable if and only if $S(x^*, F)(\omega)$ is F-measurable.

Lemma 4: ([16] Theorem 1.7.7) If (Ω, F) is a separable space, X,Y are separable metric space $F: \Omega \times X \to K(Y)$ satisfy:

(a) for any $x \in X$, $\omega \to F(\omega,x)$ is measurable;

(b) for any $\omega \in \Omega$, $x \to F(\omega,x)$ is continuous or is continuous with respect to Hausdorff metric Then $(\omega,x) \to F(\omega,x)$ is jointly measurable.

Then by Lemma 1 we have the following:

Lemma 5: Assume $F \in M^2\left([0,T] \times \Omega, \Sigma, K_{kc}(R^d); v_A\right)$. Then $S(x^*, F(t,\omega)): [0,T] \times \Omega \to R$ is Σ-measurable.

Theorem 2: Assume $F \in M^2\left([0,T] \times \Omega, \Sigma, K_{kc}(R^d); v_A\right)$. Then $I_t(F) \in L^2(\Omega; R^d)$ for each $t \in [0,T]$. Furthermore, the mapping $\psi(t,\omega) = I_t(F)$ is $(B[0,t] \otimes F_t)$-measurable.

Proof 2: Step 1: We will show that $I_t(F)$ is F_t-measurable for each $t \in [0,T]$, $\psi(t,\omega) = I_t(F)$ is $(B[0,T] \otimes F_T)$-measurable.

By Theorem 1, we have

$$I_T(F)(\omega) = \int_0^T F_t(\omega) dA_t(\omega)$$

$$= \left\{ \int_0^T f_t(\omega) dA_t(\omega); f \in S^2(F) \right\} \in K_{kc}(R^d) \tag{6}$$

for all $\omega \in \Omega$. Furthermore, we obtain

$$S(x^*, I_t(F)(\omega)) = \int_0^t S(x^*, F(s,\omega)) dA_s$$

for all $\omega \in R^d$. Moreover, since $F(t,\omega): [0,t] \times \Omega \to K_k(R^d)$ is $(B[0,t] \otimes F_t)$-measurable, from the Lemma 5 we can obtain that the function $S(x^*, F(s,\omega)): [0,T] \times \Omega \to R$ is $(B([0,T]) \otimes F_T)$ measurable.

Set-Valued Stochastic Integrals with Respect to Finite Variation

By Fubini theorem, $\int_0^t S(x^*, F(s,\omega))dA_s$ is F_t-measurable, based on Lemma 3, $I_t(F)$ is F_T-measurable.

Finally, in the argument above, the function $\int_0^t S(x^*, F(s,\omega))dA_s$ is F_T-measurable for each $t \in [0,T]$. Since it is continuous in $t \in [0,T]$ for all $\omega \in \Omega$, so it is $(B([0,T]) \otimes F_T)$-measurable. From Lemma 4, we obtain that $I_t(F)(\omega)$ is $(B([0,t]) \otimes F_t)$-measurable.

Step 2: In this step, we will show that $I_t(F)(\omega) \in L^2(\Omega; K(R^d))$ for each $t \in [0,T]$.

For each $\omega \in \Omega, t \in [0,T]$ and $f \in S^2(F)(\omega)$, we have

$$\left\| \int_0^t f_s dA_s \right\|^2 \le \mu_A([0,T]) \int_0^t \|f_s\|^2 |dA_s|$$

$$\le \mu_A([0,T]) \int_0^t \|F_s\|_k^2 |dA_s| \qquad (7)$$

then

$$\sup_{f \in S^2(F)} \left\| \int_0^t f_s dA_s \right\|^2 \le \mu_A([0,T]) \int_0^t \|F_s\|_k^2 |dA_s|$$

Hence,

$$E\left[\|I_t(F)(\omega)\|_k^2 \right]$$

$$\le E\left[\int_0^T \mu_A([0,T]) \|F(t,\omega)\|_k^2 |dA_s| \right]$$

$$\leq \int_{\Omega \times [0,T]} \|F_s\|_k^2 \, v_A(ds) < \infty, \tag{8}$$

which implies

$$I_t(F) \in L^2\left(\Omega; \mathcal{K}(R^d)\right).$$

As a manner similar to Theorem 3.8. in [8], we have the Castaing representation as following:

Theorem 3: For a set-valued stochastic process $F \in M^2\left([0,T] \times \Omega, \Sigma, K_{kc}(R^d); v_A\right)$, there exists a sequence $\{f^n; n \in \mathbb{N}\} \subset S^2(F)$ such that

$$F_t(\omega) = \mathrm{cl}\{f_t^n(\omega); n \in \mathbb{N}\} \text{ for } a.e. \, (t, \omega),$$

and, for $0 \leq s \leq t \leq T$,

$$I_{s,t}(F)(\omega) = \mathrm{cl}\left\{\int_s^t f_u^n(\omega) \, dA_u; n \in \mathbb{N}\right\} \, a.s,$$

where cl denotes the closure in R^d.

Theorem 4: For each $F \in M^2\left([0,T] \times \Omega; K_{kc}(R^d); v_A\right)$, $I_t(F)(\omega)$ is continuous a.s. with respect to the Hausdorff metric d_H.

Proof 3: Let $0 \leq r < t \leq T$ and $\omega \in \Omega$. We then have

$$I_T\left(1_{[0,t]} F\right)(\omega) = I_T\left(1_{[0,r]} F + 1_{[r,t]} F\right)(\omega)$$
$$= I_T\left(1_{[0,r]} F\right)(\omega) + I_T\left(1_{[r,t]} F\right)(\omega) \tag{9}$$

Set-Valued Stochastic Integrals with Respect to Finite Variation

Hence,

$$d_H(I_t(F)(\omega), I_r(F)(\omega))$$

$$= d_H\left(I_T\left(1_{[0,r]}F\right)(\omega) + I_T\left(1_{[r,t]}F\right)(\omega), I_T\left(1_{[0,r]}(F)(\omega)\right)\right)$$

$$\leq d_H\left(I_T\left(1_{[r,t]}F(\omega), \{0\}\right)\right)$$

$$= \sup_{f \in S_T(F)(\omega)} \left\|\int_0^t f(s)\, dA_s\right\|$$

$$\leq \int_r^t \|F(s,\omega)\|_k\, |dA_s| < \infty \tag{10}$$

since for each $f \in S_T(F)(\omega)$ $\left\|\int_0^t f(s)\,dA_s\right\| \leq \int_r^t \|F(s,\omega)\|_k\, |dA_s|$.

Hence

$$\lim_{r \uparrow t} \int_r^t \|F(s,\omega)\|_K\, dA_s(\omega) = 0$$

So $I_t(F)(\omega)$ is leftcontinuous in $t \in [0,T]$ for all a.s. In a similar way, we see that $I_t(F)(\omega)$ is right-continuous in $t \in [0,T]$ a.s.

Similar to the proof of Theorem 3.15 in [8], we have the following theorem:

Theorem 5: Let $F, G \in M^2\left(\Omega \times [0,T], \Sigma, v_A; K_{kc}(R^d)\right)$, for any $t \in [0,T]$, we have

$$E\left[d_H(I_t(F), I_t(G))\right] \leq \int_{\Omega \times [0,T]} d_H(F_s, G_s)\, v_A(ds)$$

and

$$E\left[d_H^2(I_t(F), I_t(G))\right] \leq 2\int_{\Omega \times [0,T]} d_H^2(F_s, G_s)\, v_A(ds).$$

CONCLUSIONS

When the integrand takes values in compact and convex subsets of R^d, we defined the integral with respect to real-valued variation processes. And then we proved some properties of this kind of integral such as measurability, L^2-boundedness and continuity under the Hausdorff metric.

ACKNOWLEDGEMENTS

We would like to thank the referees for their valuable comments. Moreover, we express special thanks to our editor of the journal APM for his(her) efficiency and support.

REFERENCES

1. M. Kisielewicz, "Set-Valued Stochastic Integrals and Stochastic Inclusions," Discussiones Mathematicae, Vol. 13, 1993, pp. 119-126.
2. B. K. Kim and J. H. Kim, "Stochastic Integrals of Set-Valued Processes and Fuzzy Processes," Journal of Mathematical Analysis and Applications, Vol. 236, No. 2, 1999, pp. 480-502. http://dx.doi.org/10.1006/jmaa.1999.6461
3. R. J. Aumann, "Intgrals of Set-Valued Functions," Journal of Mathematical Analysis Applications, Vol. 12, No. 1, 1965, pp. 1-12. http://dx.doi.org/10.1016/0022-247X(65)90049-1
4. E. J. Jung and J. H. Kim, "On Set-Valued Stochastic Integrals," Stochastic Analysis and Applications, Vol. 21, No. 2, 2003, pp. 401-408. http://dx.doi.org/10.1081/SAP-120019292
5. S. Li and A. Ren, "Representation Theorems, Set-Valued and Fuzzy Set-Valued Ito Integral," Fuzzy Sets and Systems, Vol. 158, No. 9, 2007, pp. 949-962. http://dx.doi.org/10.1016/j.fss.2006.12.004
6. J. Zhang, "Set-Valued Stochastic Integrals with Respect to a Real Valued Maringale," Soft Method for Handling Vaelability and Imprecision ASC 48, Spinger-Verlag, Berlin Herdelberg, 2008.
7. J. Zhang, S. Li, I. Mitoma and Y. Okazaki, "On Set-Valued Stochastic Integrals in M-Type 2 Banach Space," Journal of Mathematical Analysis and Applications, Vol. 350, No. 1, 2009, pp. 216-233. http://dx.doi.org/10.1016/j.jmaa.2008.09.017
8. J. Zhang, S. Li, I. Mitoma and Y. Okazaki, "On the Solution of Set-Valued Stochastic Differential Equations in M-Type 2 Banach Space," Tohoku Mathematical Journal, Vol. 61, No. 3, 2009, pp. 417-440. http://dx.doi.org/10.2748/tmj/1255700202

9. J. Zhang, "Integrals and Stochastic Differential Equations for Set-Valued Stochastic Processes," Ph.D. Thesis, Saga University, Saga, 2009.
10. I. Mitoma, Y. Okazaki and J. Zhang, "Set-Valued Stochastic Differential Equations in M-Type 2 Banach Space," Communications on Stochastic Analysis, Vol. 4, No. 2, 2010, pp. 215-237.
11. J. G. Li, S. Li and Y. Ogura, "Strong Solutions of Ito Type Set-Valued Stochastic Differential Equation," Acta Mathematica Sinica, English Series, Vol. 26, No. 9, 2010, pp. 1739-1748. http://dx.doi.org/10.1007/s10114-010-8298-x
12. M. Malinowski and M. Michta, "Set-Valued Stochastic Integral Equations Driven by Martingales," Journal of Mathematical Analysis and Applications, Vol. 394, No. 12, 2012, pp. 30-47. http://dx.doi.org/10.1016/j.jmaa.2012.04.042
13. J. Zhang, I. Mitoma and Y. Okazaki, "Set-Valued Stochastic Integrals with Respect to Poisson Processes in a Banach Space," International Journal of Approximate Reasoning, Vol. 54, No. 3, 2013, pp. 404-417. http://dx.doi.org/10.1016/j.ijar.2012.06.001
14. Z. Wang and R. Wang, "Set-Valued Lebesgue-Stieltjes Integrals," Journal of Applied Probability and Statistics, Vol. 13, No. 3, 1997, pp. 303-316.
15. S. Li, Y. Ogura and Y. Kreinovich, "Limit Theorems and Applications of Set-Valued and Fuzzy Set-Valued Random Variables," 43rd Edition, Kluwer Academic Publishers, Dordrecht, 2002. http://dx.doi.org/10.1007/978-94-015-9932-0
16. W. Zhang, S. Li, Z. Wang and Y. Gao, "An Introduction about Set-Valued Stochastic Process," Science Press, Beijing, 2007.
17. L. Wang and H. Xue, "Set-Valued Lebesgue-Stieltjes Integrals," Basic Sciences Journal of Textile Universities, Vol. 16, No. 4, 2004, pp. 317-320.

CITATION

J. Zhang and J. Qi, "Set-Valued Stochastic Integrals with Respect to Finite Variation Processes," *Advances in Pure Mathematics*, Vol. 3 No. 9A, 2013, pp. 15-19. doi: 10.4236/apm.2013.39A1003.

A New Formula For Partitions In A Set Of Entities Into Empty And Nonempty Subsets, And Its Application To Stochastic And Agent-Based Computational Models

Ghennadii Gubceac, Roman Gutu, Florentin Paladi
Department of Theoretical Physics, State University of Moldova, Chisinau, Republic of Moldova

ABSTRACT

In combinatorics, a Stirling number of the second kind S(n,k) is the number of ways to partition a set of n objects into k nonempty subsets. The empty subsets are also added in the models presented in the article in order to describe properly the absence of the corresponding type i of state in the system, i.e. when its "share" $p_i=0$. Accordingly, a new equation for partitions P(N,m) in a set of entities into both empty and nonempty subsets was derived. The indistinguishableness of particles (N identical atoms or molecules) makes only sense within a cluster (subset) with the size $0 \leq n_i \leq N$. The first-order phase transition is indeed the case of transitions, for example in the simplest interpretation, from completely liquid state $typeL = \{n_1 = N, n_2 = 0\}$ to the completely crystalline state $typeC = \{n_1 = N, n_2 = N\}$. These partitions are well distinguished from the physical point of view, so they are 'typed' differently in the model. Finally, the present developments in the physics of complex systems, in particular the structural relaxation of supercooled liquids and glasses, are discussed by using such stochastic cluster-based models.

INTRODUCTION

In mathematics, particularly in combinatorics and the study of partitions, a Stirling numbers of the second kind $S(n,k) \equiv \{{n \atop k}\}$ count the number of ways to partition a set of n labeled objects into k nonempty unlabelled subsets. By the way, Stirling numbers of the second kind show up more often than those of first and third kind (or Lah numbers), and James Stirling himself considered this kind first [1]. Equivalently, they count the number of different equivalence relations with precisely k equivalence classes that can be defined on an n element set, and they can be calculated using the explicit formula $S(n,k) = \frac{1}{k!}\sum_{j=0}^{k}(-1)^{k-j}\binom{k}{j}j^n$, where $\binom{k}{j}$ is the binomial coefficient. The sum over the values for k of the Stirling numbers of the second kind, i.e. $\sum_{k=0}^{k}\{{n \atop k}\} = B_n$ gives the nth Bell number, that is the total number of partitions of a cluster with n entities or agents [2]. The question arises when the empty subsets or clusters are needed to be used in the models, for example, with heterogeneous structure interactions and, subsequently, the partition by Stirling numbers of the second kind becomes inappropriate to count partitions in such complex systems [3].

In general, agent-based modeling is currently a technique widely used to simulate complex systems in computer science and social sciences. On the other hand, a Markovian process is a stochastic process whose future probabilities are determined by its most recent values. The agent-based computational models (ABM) fits well this description, except for the cases when decisions are dependent on the state of the systems of more than one steps ago, which is the case when ABM agents experience learning, adaptation, and reproduction [4].

In this study, a general equation that describes clustering process among interacting agents in heterogeneous populations, i.e. the partition process in a set of entities into empty and nonempty clusters, is derived and used to study how different behavioral norms affect the individual and social welfare in a population with heterogeneous preferences.

One can consider them as an idealization of an imperative and a more liberal approach to social norms or stylized behavioral rules studied by agent-based computational models [5,6]. Another application refers to the generic stochastic model for crystal nucleation which is useful to depict the impact of interface between the nucleus considered as a cluster of a certain number of molecules and the liquid phase for the enhancement of the overall nucleation process. It is generally known that first-order phase transitions occur by nucleation mechanism, and both the nucleus, a cluster of atoms or molecules, and the nucleation work, a energy barrier to the phase transition, are basic thermodynamic quantities in the theory of nucleation. However, the critical nucleus formation is statistically a random event with a probability largely determined by the nucleation work which increases with the subnuclei size [7]. The traditional differential equation modeling is not the alternative to agent-based models; only a set of differential equations, each describing the dynamics of one of the system's constituent units, is an agent-based model [8].

The general formulation is outlined in Section 2. In Section 3 a probabilistic approach to the crystal nucleation process is considered. The main conclusions are presented in Section 4.

THE MODEL

There are N entities which can be in 3 different states (call them left, center and right), and can play 3 actions (again left, center and right). Interaction in this agentbased model involves always one active and one passive player, but agents can play both roles interchangeably. They have preferences over their states: accept one state, are neutral with respect to another state and reject the remaining state. When two agents meet, the active player sets the passive player's state according to his actionwhich in turn is determined by one of the applied rule. This identifies only 6 possible combinations. Denote with $p_1 \cdots p_6$ the shares of the population characterized by each combination of preferences, as in Table 1. That is, drawing randomly one agent, it will be of type i with probability p_i. After each interaction, the passive player gets a payoff of +1 if it is in the accepted state, a payoff of 0 if it is in the neutral state, and a payoff of −1 if it is in the rejected state. The active player does not get any feedback. If the active player follows the first J-rule, it always plays the action corresponding to the accepted state. If it follows the second H-rule, it randomizes between actions corresponding to the accepted and neutral states. The further example will clarify. Suppose two individuals, A and B, meet. Player A is the active one, and rejects left, accepts right, and is thus neutral with respect to the center. Player B is the passive one. It accepts left, rejects right, and is neutral with respect to center, like player A. Suppose A follows the J-rule, and will play right, setting B's state to right. B will then have a payoff of −1. Suppose, on the other hand, that A follows the H-rule, and will randomize between center and right. The payoff for B could then be either 0 or −1. Note that there is no strategic interaction in the model: the passive player's payoff depends on the active player's choice, but the active player's choice does not depend on the passive player in any way, so the game-theoretic solution concepts like Nash equilibrium become useless.

Aggregate results are defined in terms of both the mean π and the variance σ^2 of the payoffs which denote the stability and the heterogeneity of population, respectively. However, in order to avoid arbitrary choices we do not specify a particular functional form, and report separately the results for the mean and the variance. Let $N = 1, 2, \cdots, \infty$ be the total number of entities in the model, and the $\{n_1, n_2, n_3, n_4, n_5, n_6\}$ is their partition into $m = 6$ subsets. Each subset can be called cluster, and the process itself—clustering. The size of each cluster can vary from 0 to N, $n_i = \overline{0, N}, i = \overline{1, 6}$, and $\sum_{i=1}^{6} n_i = N$. The number of possible partitions P is a

A New Formula For Partitions In A Set Of Entities Into Empty

Table 1: Distribution of preferences in the heterogeneous system with 6 types of states.

Type	Accepted State	Rejected State	Share	left	center	right
1	left	center	p_1	p_1	$-p_1$	0
2	left	right	p_2	p_2	0	$-p_2$
3	center	left	p_3	$-p_3$	p_3	0
4	center	right	p_4	0	p_4	$-p_4$
5	right	left	p_5	$-p_5$	0	p_5
6	right	center	p_6	0	$-p_6$	p_6
Coordinates				$p_1+p_2-p_3-p_5$	$-p_1+p_3+p_4-p_6$	$-p_2-p_4+p_5+p_6$

function of N and m, and the explicit solution is

$$P(N, m=6)$$

$$= {}_6C_1 + (N-1)\,{}_6C_2 + \left(1 + \sum_{i=2}^{N-2} i\right){}_6C_3$$

$$+ \left(N - 3 + (N-2)\sum_{i=2}^{N-3} i - \sum_{i=2}^{N-3} i^2\right){}_6C_4$$

$$+ \left(1 + \sum_{i=2}^{N-4} i + \sum_{i=3}^{N-3} i \sum_{l=2}^{i-1} l - (N-3)\sum_{i=2}^{N-4} i^2 + \sum_{i=2}^{N-4} i^3\right){}_6C_5$$

$$+ (N-5)\sum_{i=2}^{N-5}(N-i-4)i + \sum_{i=3}^{N-4}(N-i-3)i\sum_{l=2}^{i-1} l$$

$$- \sum_{i=2}^{N-5} i^2 \sum_{l=i+1}^{N-4} l + \left(\sum_{i=2}^{N-5}(N-i-4)i^3\right){}_6C_6$$

$$= 2(74N - 213) + 2(11N - 13)\sum_{i=2}^{N-5} i$$

$$+ 2(1 - 3N)\sum_{i=2}^{N-5} i^2 + (2+N)\sum_{i=2}^{N-5} i^3 - \sum_{i=2}^{N-5} i^4$$

$$+ \sum_{i=3}^{N-4}\left((3+N)i - i^2\right)\sum_{l=2}^{i-1} l - \sum_{i=2}^{N-5} i^2 \sum_{l=i+1}^{N-4} l$$

$$= \frac{1}{5!}\prod_{i=1}^{5}(N+i),$$

where $_6C_1 =\, _6C_5 = 6, _6C_2 =\, _6C_4 = 15, =\, _6C_3 = 20, =\, _6C_6 = 1$ are combinations given by the binomial coefficient $_mC_k \equiv \binom{m}{k} \equiv \frac{m!}{(m-n)!k!}$. In general, when the number of subsets m>1, the number of possible partitions P of N agents into m subgroups or subsets is

$$P(N,m) = \frac{1}{(m-1)!}\prod_{i=1}^{m-1}(N+i). \tag{1}$$

The matrix is diagonally symmetric

$$P(N, i+1) = P(i, N+1)$$

for i = 0,1,···,N−1, and it can be formed by arranging the partition numbers according to the parameters N and m. Table 2 contains this array of values for the given numbers of partitions. The reccurence relation is

$$P(N,m) = P(N-1,m) + P(N,m-1)$$

for m>0 with the initial condition $P(0, m > 0) = 1$. For instance, the number 330 in column m=5 and row N=7 is given by $330 = 210 + 120$, where 210 is the number above and 120 is the number on the left of 330. The diagonal elements of the bi-triangular array of values for the numbers of partitions are 1, 2, 6, 20, 70, 252, 924, 3432, 12,870, 48,620, ..., and it is represented by nth central binomial coefficient:

$$C(2n,n) \equiv \binom{2n}{n} = \frac{(2n)!}{(n!)^2} \quad \text{for all} \quad n \geq 0.$$

They are called central since they show up exactly in the middle of the even-numbered rows in Pascal's traingle. These numbers have the generating function

A New Formula For Partitions In A Set Of Entities Into Empty

$$\frac{1}{\sqrt{1-4x}} = 1 + 2x + 6x^2 + 20x^3 + 70x^4 + 252x^5$$
$$+ 924x^6 + 3432x^7 + \cdots.$$

It is known that the asymptotics for the central binomial coefficient $C(2n,n)$ can be written in the form of a particular case of the Wallis formula, i.e.

$$\lim_{n \to \infty} \frac{2^{4n}}{n\binom{2n}{n}^2} = \pi \lim_{n \to \infty} \frac{n\left[\Gamma(n)\right]^2}{\left[\Gamma\left(\frac{1}{2}+n\right)\right]^2} = \pi.$$

Table 2: Bi-triangular array of values for the numbers of partitions P(N,m).

N,m	1	2	3	4	5	6	7	8	9	10	Recurrence Relation
0	1	1	1	1	1	1	1	1	1	1	P(N,m) =
1	1	2	3	4	5	6	7	8	9	10	P(N−1,m) +
2	1	3	6	10	15	21	28	36	45	55	P(N,m−1)
3	1	4	10	20	35	56	84	120	165	220	
4	1	5	15	35	70	126	210	330	495	715	
5	1	6	21	56	126	252	462	792	1287	2002	
6	1	7	28	84	210	462	924	1716	3003	5005	
7	1	8	36	120	330	792	1716	3432	6435	11,440	
8	1	9	45	165	495	1287	3003	6435	12,870	24,310	
9	1	10	55	220	715	2002	5005	11,440	24,310	48,620	
10	1	11	66	286	1001	3003	8008	19,448	43,758	92,378	

where $\Gamma(x)$ is the gamma function, so

$$\binom{2n}{n} \sim \frac{4^n}{\sqrt{\pi n}}, \quad n \to \infty.$$

By the way, this equation can also be used to determine the constant $\sqrt{2\pi}$ in front of the Stirling's formula.

Now it is straightforward to see that when all individuals share the same preferences (polarization) the first rule gives a higher payoff. In the other extreme case, when preferences are equally distributed in the ensemble (dispersion) and $p_1 = p_2 = \cdots = p_6 = \dfrac{1}{6}$, it is again straightforward to see that the rules are equivalent, and lead to an average payoff $\pi = 0$. Consider next an active player of type 1 (accepts left and rejects center) which meets in turn all other (passive) agents, including himself. If it follows the first rule, then it will play left causing a payoff of +1 in $(p_1 + p_2)N$ agents, and a payoff of -1 in $(p_3 + p_5)N$ agents. Note that there are $(p_1 + p_2)N$ similar entities in the ensemble. Suppose now that everybody meets everybody else both as active and as passive players. Coupling them randomly and randomly choosing active and passive players only adds some noise to the simulation results. So the average payoff when everybody plays according to the first J-rule is

$$\pi_1 = (p_1 + p_2)(p_1 + p_2 - p_3 - p_5)$$
$$+ (p_3 + p_4)(-p_1 + p_3 + p_4 - p_6)$$
$$+ (p_5 + p_6)(-p_2 - p_4 + p_5 + p_6). \tag{2}$$

Similarly, the average payoff with the second H-rule is

$$\pi_2 = \frac{1}{2}\big(p_1(p_1 - p_3 - p_5 + p_6)$$
$$+ p_2(p_2 - p_3 + p_4 - p_5) + p_3(-p_1 + p_3 + p_5 - p_6)$$
$$+ p_4(-p_1 + p_2 + p_4 - p_6) + p_5(-p_2 + p_3 - p_4 + p_5$$
$$+ p_6(p_1 - p_2 - p_4 + p_6)\big). \tag{3}$$

A New Formula For Partitions In A Set Of Entities Into Empty

Table 3: Expressions of $\pi_1 - \pi_2$ for different numbers of non-zero probabilities p_i, i = 1, 2, ···, 6.

No. p_i	Non-zero Probabilities	Expressions for $\delta = \pi_1 - \pi_2$
2	(p_1,p_2) (p_3,p_4) (p_5,p_6)	$\delta_1 = \frac{1}{2} + p_i - p_i^2$
	(p_1,p_3) (p_1,p_6) (p_2,p_4) (p_2,p_5) (p_3,p_5) (p_4,p_6)	$\delta_2 = \frac{1}{2}(1 - 2p_i)^2$
	(p_1,p_4) (p_1,p_5) (p_2,p_3) (p_2,p_6) (p_3,p_6) (p_4,p_5)	$\delta_3 = \frac{1}{2}(1 + 3p_i(p_i - 1))$
3	(p_1,p_2,p_3) (p_1,p_2,p_4) (p_1,p_2,p_5) (p_1,p_2,p_6) (p_1,p_3,p_4) (p_2,p_3,p_4) (p_3,p_4,p_5) (p_3,p_4,p_6) (p_1,p_5,p_6) (p_2,p_5,p_6) (p_3,p_5,p_6) (p_4,p_5,p_6)	$\delta_1 = \frac{1}{2}\left(p_h^2 + p_i^2 + p_j^2 + 4p_h p_i - 2p_h p_j - p_i p_j\right)$
	(p_1,p_3,p_5) (p_1,p_3,p_6) (p_2,p_3,p_5) (p_2,p_4,p_5) (p_1,p_4,p_6) (p_2,p_4,p_6)	$\delta_2 = \frac{1}{2}\left(p_h^2 + p_i^2 + p_j^2 - 2(p_h p_i + p_i p_j) - p_h p_j\right)$
	(p_2,p_3,p_6) (p_1,p_4,p_5)	$\delta_3 = \frac{1}{2}\left(p_h^2 + p_i^2 + p_j^2 - p_h p_i - p_h p_j - p_i p_j\right)$
4	(p_1,p_2,p_3,p_4) (p_1,p_2,p_5,p_6) (p_3,p_4,p_5,p_6)	$\delta_1 = \frac{1}{2}\left(p_h^2 + p_i^2 + p_j^2 + p_k^2 + 4(p_h p_i + p_j p_k) - 2(p_h p_j + p_i p_k) - p_i p_j - p_h p_k\right)$
	(p_1,p_2,p_3,p_5) (p_1,p_2,p_4,p_6) (p_1,p_3,p_4,p_6) (p_1,p_3,p_5,p_6) (p_2,p_3,p_4,p_5) (p_2,p_4,p_5,p_6)	$\delta_2 = \frac{1}{2}\left(p_h^2 + p_i^2 + p_j^2 + p_k^2 + 4p_h p_i - 2(p_h p_j + p_i p_k + p_j p_k) - p_i p_j - p_i p_k\right)$
	(p_1,p_2,p_3,p_6) (p_1,p_2,p_4,p_5) (p_1,p_3,p_4,p_5) (p_2,p_3,p_4,p_6) $(p_2,p3,p_5,p_6)$ (p_1,p_4,p_5,p_6)	$\delta_3 = \frac{1}{2}\left(p_h^2 + p_i^2 + p_j^2 + p_k^2 + 4p_h p_i - 2(p_h p_j + p_i p_k) - p_i p_j - p_i p_k - p_j p_k\right)$
		$\delta = \frac{1}{2}\begin{pmatrix} p_h^2 + p_i^2 + p_j^2 + p_k^2 + p_l^2 + 4(p_h p_i + p_j p_k) - 2(p_h p_j + p_i p_k) \\ -2(p_h p_l + p_i p_k + p_i p_l + p_j p_l) - p_i p_j - p_h p_k - p_h p_l - p_k p_l \end{pmatrix}$

To study the behavior of $\pi_1 - \pi_2$ based on Equations (2) and (3), it is convenient to set some of the probabilities to zero. It is straightforward to see that when there is just one probability different from 0 (and thus equal to 1) we have $\pi_1 - \pi_2 = \frac{1}{2} > 0$. In general, the number of different possible combinations of non-zero probabilities is given by the above binomial coefficient ${}_mC_{kl}$.

Thus for two non-zero probabilities we obtain a set of ${}_6C_2 = 15$ equations which can be grouped in just 3 different functional forms shown in Table 3. Figure 1 graphs these 3 curves for all values of p_i and $p_j = 1 - p_i$. The J-rule still performs better in all cases but one, when

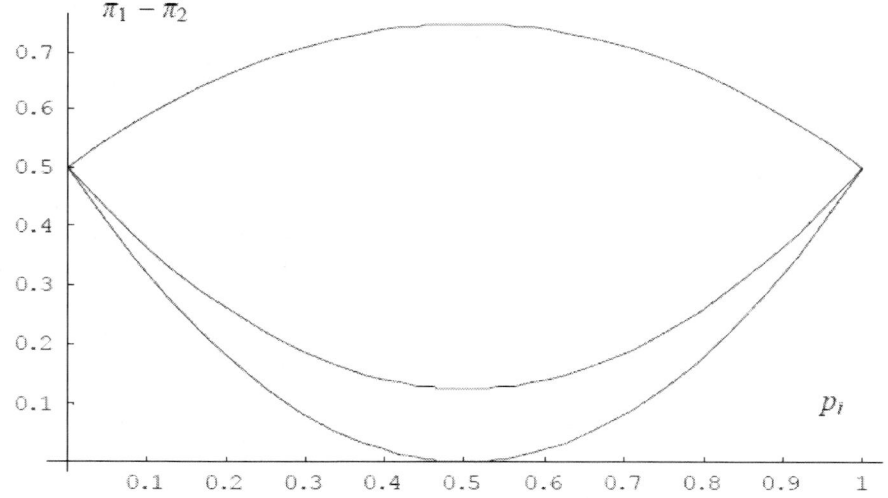

Figure 1: $\pi_1 - \pi_2$ in the case of two non-zero probabilities, p_i and $p_j = 1 - p_i$.

the two rules are equivalent. However, from three nonzero probabilities onward things start to look differently. For three non-zero probabilities we have $_6C_3 = 20$ equations, while for four non-zero probabilities there are $_6C_4 = 15$ equations, and for five non-zero probabilities $_6C_5 = 6$ equations. These equations reduce to just three different functional forms in case of three and four nonzero probabilities, and to just one expression in case of five non-zero probabilities.

A trend towards a better performance of the H-rule, as the distribution of preferences in the system becomes less polarized, is evident. However, in order to better investigate it, a definition of how much preferences are polarized is needed. We represent the distribution of states as a single point in a three dimensional space, where the axes are labeled l, c and r. The l coordinate is found by counting all agents who accept left, and subtracting all agents who reject left. The result is then normalized to the size of the population. Similarly for the other two coordinates. Hence,

$$l = p_1 + p_2 - p_3 - p_5,$$
$$c = p_3 + p_4 - p_1 - p_6,$$
$$r = p_5 + p_6 - p_2 - p_4, \qquad (4)$$

Where $l + c + r = 0$. Note that different distributions of states can lead to the same point in the sphere. For instance, the point in the origin is given not only by $p_1 = p_2 = \cdots p_6 = 1/6$, but by any combination of preferences such as $p_1 = p_3, p_2 = p_5,$ and $p_4 = p_6$. Taking into account Equation (4), one can define now the polarization of states as the distance from the center of the sphere:

$$d(l,r,c) \equiv d(p_1, p_2, \cdots, p_6) = \sqrt{l^2 + r^2 + c^2} \; . \; . \qquad (5)$$

Note that $d \in \left[0, \sqrt{2}\right]$: all points thus lie inside a sphere around the origin.

The variances σ_1^2 and σ_2^2 are defined for each discrete distribution $D \equiv 1,2$ with the expectation (mean) value π_D as follows:

$$\sigma_D^2 = \sum_{i=1}^{6} p_i \left(\pi_{i,D} - \pi_D\right)^2, \tag{6.1}$$

where

$$\pi_{1,1} = p_1 + p_2 - p_3 - p_4,$$
$$\pi_{4,1} = p_3 + p_4 - p_5 - p_6,$$
$$\pi_{2,1} = p_1 + p_2 - p_5 - p_6,$$
$$\pi_{5,1} = -p_1 - p_2 + p_5 + p_6,$$
$$\pi_{3,1} = -p_1 - p_2 + p_3 + p_4,$$
$$\pi_{6,1} = -p_3 - p_4 + p_5 + p_6 \tag{6.2}$$

and

$$\pi_{1,2} = (p_1 - p_3 - p_5 + p_6)/2,$$
$$\pi_{4,2} = (-p_1 + p_2 + p_4 - p_6)/2,$$
$$\pi_{2,2} = (p_2 - p_3 + p_4 - p_5)/2,$$
$$\pi_{5,2} = (-p_2 + p_3 - p_4 + p_5)/2,$$
$$\pi_{3,2} = (-p_1 + p_3 + p_5 - p_6)/2,$$
$$\pi_{6,2} = (p_1 - p_2 - p_4 + p_6)/2. \tag{6.3}$$

Figure 2 explores how outcomes vary as functions of the distance $d(l,r,c)$ defined by Equation (5). The whole range $[0,1]$ is sampled, for all probabilities $p_1 \cdots p_6$. The step considered for creating all com-

A New Formula For Partitions In A Set Of Entities Into Empty

binations of probabilities is 0.025, i.e. the total number of agents is 40. The average values for π_1 and π_2 are shown in Figure 2(a), and for each value of the distance from the center of the sphere, $d(l,r,c)$, the frequency of wins with each rule is computed (see Figure 2(b)). When $\pi_1 - \pi_2 > 0$ a win is assigned to the first J-rule, and when $\pi_1 - \pi_2 < 0$ a win is assigned to the second

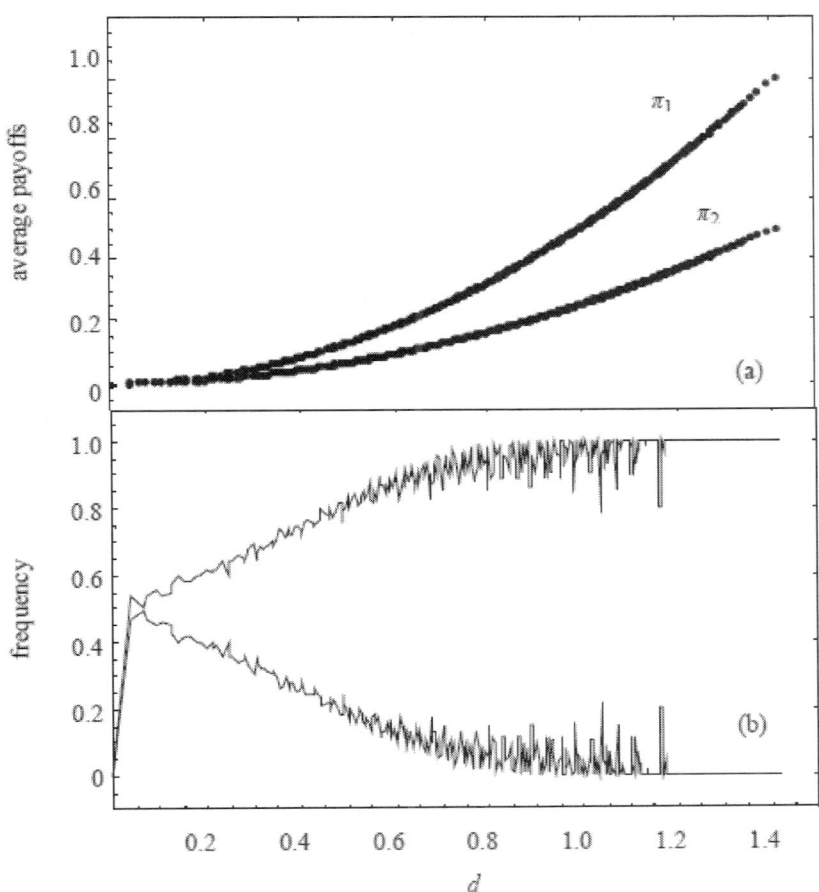

Figure 2: Average values for π_1 (upper line) and π_2 (a), and relative frequency of negative and positive values for $\pi_1 - \pi_2$ (b) as functions of the distance $d(p_1, p_2, \cdots p_6)$ from the center of sphere with a radius of $\sqrt{2}$: 40 entities, 6 clusters in a similar ABM model.

one. Exactly in the center of the sphere the two rules lead to the same payoff, independently of the underlying distribution of states. Close to the center, each rule wins in about 50% of the cases. Then, as we move away from the center, the first rule improves its performance, and it is always better when the states are totally polarized, but the total number of states for intermediate values of the distance d is much larger than for the dispersed and polarized states.

Meanwhile, the variance (6.1) with the J-rule (6.2) is generally higher than the variance with the H-rule (6.3), especially when the preferences are dispersed in the system, and when they are quite polarized. On average, however, when one rule is better in terms of higher expected payoffs it is also better in terms of lower heterogeneity, and there are the same average values of the scale-free coefficients of variation on the distance d for any level of fragmentation of states.

A PROBABILISTIC APPROACH TO THE CRYSTAL NUCLEATION PROCESS

The second application refers to the nucleation process, a widely spread phenomenon in both nature and technology, which may be considered as a representative of the aggregation phenomena in complex systems. Let's consider N atoms which can be in 3 different states (cluster, liquid and their interface), and can perform 4 possible moves: liquid to interface, interface to liquid, interface to cluster, and cluster to interface. One can identify 4 different combinations denoted with probabilities $p_1 \cdots p_4$, as in Table 4. That is, drawing randomly one particle, it will be of type i with probability p_i. Let $N = 1, 2, \cdots, \infty$ be the total number of atoms in the system, and $\{n_1, n_2, n_3, n_4\}$ are their partition into 4 subsets. Each subset can be called cluster, and the number of possible partitions (1) in this case is

$$P(N, m = 4) = \frac{1}{3!} \prod_{i=1}^{3} (N + i),$$

where $n_i = \overline{0,N}, i = \overline{1,4}$ and $\sum_{i=1}^{4} n_i = N$. For example, in a system of $N=1000$ atoms, $P(N, m = 4)$ equals to 167,668,501! Accordingly, the number of repeated computer runs in an ABM model would be very large due to different possible partitions. But we are able to overcome this problem, as already mentioned in the previous section, by developing similar stochastic mathematical models which can describe exactly the results of the agentbased computational models, and, finally, by bridging the gap between ABM modeling and stochastic processes.

Let's consider further that each particle interacts with the entire group both as an aggressor in terms of the Kolmogorov mathematical theory [9], and as a passive agent in terms of the ABM computational models as well. Then the mean π, namely the stability index, takes here the form

$$\pi = p_1(p_1 - p_3) + (p_2 + p_3)(-p_1 + p_2 + p_3 - p_4)$$
$$+ p_4(p_4 - p_2) \tag{7.1}$$

or, taking into account that $\sum_{i=1}^{4} p_i = 1$, one can exclude one probability, for example p_4, from the above equation:

$$\pi = p_1(p_1 - p_3)$$
$$+ (1 - p_1 - 2p_2 - p_3)(1 - p_1 - p_2 - p_3)$$
$$+ (p_2 + p_3)(-1 + 2p_2 + 2p_3). \tag{7.2}$$

One can represent again the distribution of states as a three-dimensional point (l,r,c) inside a sphere, and the mean payoff π (Equation (7.1) or Equation (7.2)) can be obtained as a function of the distance

$$d(l,r,c) \equiv d(p_1, p_2, p_3, p_4) = \sqrt{l^2 + r^2 + c^2},$$

where the axes are labeled l, c and t: $l = p_1 - p_3$, $c = p_2 + p_3 - p_1 - p_4$, $r = p_4 - p_2$, and, as stated in the previous section, $l + c + r = 0$ and $d \in \left[0, \sqrt{2}\right]$. Thus different distributions of states can lead to the same point in the sphere, i.e. different microscopic partitions can

Table 4: Distribution of accepted and rejected states in the system.

Type	Accepted State	Rejected State	Share	cluster	*interface*	liquid
1	cluster/nucleus	interface	p_1	p_1	$-p_1$	0
2	Interface	liquid	p_2	0	p_2	$-p_2$
3	interface	cluster/nucleus	p_3	$-p_3$	p_3	0
4	liquid	interface	p_4	0	$-p_4$	p_4
Coordinates				$p_1 - p_3$	$p_1 - p_2 + p_3 - p_4$	$-p_2 + p_4$

generate the same result on aggregate inside a sphere around the origin. The results for the two limit cases are obviously: if all particles would show the same behavior, then $d = \sqrt{2}$ and there is a maximum stability of states in such a completely asymmetrical system, but $\pi = 0$ for a homogeneous system, $p_1 = p_2 = p_3 = p_4 = 1/4$, and for combinations such as $p_1 = p_3$ and $p_2 = p_4$ in the case of unstable states.

CONCLUSIONS

Mathematical models developed in this paper were used together with agent-based computations to study interactions in two quite different heterogeneous systems. The models are general ones and allow simulations for any size of the system. Our results also support the idea that general properties observed in the complex systems of any kind, which arise from the interplay between random interactions and their complex structures, could be successfully investigated by the same tool. For example, these properties are frequently found in the physical sciences, and are the domain of non-equilibrium statistical mechanics. Because of the real limitations in the use of analytical methods to study such problems, it is often necessary to resort to numerical ones, and the advent of computer simulations has led to an increase of scientific activity in this area that has emerged nowadays as a major subject of interdisciplinary research.

The recent discussions about the gap between agentbased computational models and stochastic analytical models have stimulated the research on this topic [3,10]. In such context, the article supports a unified probabilistic approach to simulation of multi-agent interactions in heterogeneous complex systems. This method appears phenomenological if it is not agent-based, and we prove analytically that the average outcomes of multiple agentbased computations can be described precisely. Furthermore, a useful aggregation procedure for representing the three-dimensional distribution of states, and a general formula that describes clustering process among interacting agents in heterogeneous populations, which is the Equation (1) for partitions in a set of entities into empty and nonempty subsets, are proposed. Bi-triangular array of values for the numbers of partitions is presented, and its diagonal elements are represented by nth central binomial coefficient.

In particular, we obtained that different distributions of states can lead to the same point inside the sphere around the origin, i.e. different microscopic partitions can generate the same aggregate result. It would be a particular conclusion set for the analyzed situations, but one would expect this in different circumstances due to the universal aspects observed in the behavior of complex systems.

ACKNOWLEDGMENTS

F.P. thanks Dr. Matteo Richiardi for his ABM contribution.

REFERENCES

1. R. L. Graham, D. E. Knuth and O. Patashnik, "Concrete Mathematics. A Foundation for Computer Science," 2nd Edition, Addison-Wesley Professional, Reading, 1994.
2. J. Sandor and B. Crstici, "Handbook of Number Theory II," Kluwer Academic Publishers, Dordrecht, 2004. http://dx.doi.org/10.1007/1-4020-2547-5
3. F. Paladi, "On the Probabilistic Approach to Heterogeneous Structure Interactions in Agent-Based Computational Models," Applied Mathematics and Computation, Vol. 219, No. 24, 2013, pp. 11430-11437. http://dx.doi.org/10.1016/j.amc.2013.05.042
4. E. Bonabeau, "Agent-Based Modeling: Methods and Techniques for Simulating Human Systems," Proceedings of the National Academy of Sciences of the United States of America, Vol. 99, No. 3, 2002, pp. 7280-7287. http://dx.doi.org/10.1073/pnas.082080899
5. M. Richiardi and F. Paladi, "Jesus, Hillel and the Man of the Street. Moral and Social Norms in Heterogeneous Populations," LABORatorio R. Revelli Working Paper 40, 2005. http://eco83.econ.unito.it/terna/swarmfest2005papers/richiardi_paladi.pdf
6. M. Richiardi, "Jesus vs Hillel. From Moral to Social Norms and Back," European Journal of Economic and Social Systems, Vol. 19, No. 2, 2006, pp. 171-190.
7. D. Kashchiev, "Nucleation. Basic Theory with Applications," Butterworth-Heinemann, Oxford, 2000.
8. C.W. Gardiner, "Handbook of Stochastic Methods: for Physics, Chemistry and the Natural Sciences," 2nd Edition, Springer-Verlag, Berlin, 1985.
9. A. N. Kolmogorov, "On Statistical Theory of Metal Crystallisation (in Russian)," Izvestiya Academy of Sciences, USSR, Mathematics, Vol. 3, 1937, pp. 355-360.
10. L. Feng, B. Li, B. Podobnik, T. Preis and H. E. Stanley, "Linking Agent-Based Models and Stochastic Models of Financial Markets," Proceedings of the National Academy of Sciences of the United States of America, Vol. 109, No. 22, 2012, pp. 8388-8393. http://dx.doi.org/10.1073/pnas.1205013109

CITATION

G. Gubceac, R. Gutu and F. Paladi, "A New Formula for Partitions in a Set of Entities into Empty and Nonempty Subsets, and Its Application to Stochastic and Agent-Based Computational Models," *Applied Mathematics*, Vol. 4 No. 10C, 2013, pp. 14-21. doi: 10.4236/am.2013.410A3003.

Upwind Finite-Volume Solution Of Stochastic Burgers' Equation

Mohamed A. El-Beltagy[1], Mohamed I. Wafa[1], Osama H. Galal[2]

[1]Department of Engineering Mathematics & Physics, Engineering Faculty, Cairo University, Giza, Egypt

[2]Department of Engineering Mathematics & Physics, Faculty of Engineering, Fayoum University, Fayoum, Egypt

ABSTRACT

In this paper, a stochastic finite-volume solver based on polynomial chaos expansion is developed. The upwind scheme is used to avoid the numerical instabilities. The Burgers' equation subjected to deterministic boundary conditions and random viscosity is solved. The solution uncertainty is quantified for different values of viscosity. Monte-Carlo simulations are used to validate and compare the developed solver. The mean, standard deviation and the probability distribution function (p.d.f) of the stochastic Burgers' solution is quantified and the effect of some parameters is investigated. The large sparse linear system resulting from the stochastic solver is solved in parallel to enhance the performance. Also, Monte-Carlo simulations are done in parallel and the execution times are compared in both cases.

INTRODUCTION

In engineering fields, most models are represented as partial differential equations (PDEs), assuming all input data are perfectly known. Unfortunately, geometry and material characteristics for instance would rather present uncertainties. Under those conditions, the output data become also uncertain. To deal with propagation of the input data uncertainties to the output data, probabilistic models are more appropriate than deterministic ones. Several methods of solution are developed to assess the response due to the uncertainties. This response depends on two main factors: the first factor is the geometric domain discretization; the second is the discretization involved random process [1,2]. The methods of solution may be classified according to the first factor to meshless methods [3], stochastic finite difference methods [4], and stochastic finite element methods [5]. On the other hand, according to the second factor the methods of solution may be classified to Monte-Carlo simulations (MCS) [6], perturbations [7], and spectral stochastic finite element methods (SSFEM) [8,9]. Recently, the SSFEM is one of the most widely used methods [7].

On the other hand, the capability of neural network to analyze stochastic finite element is discussed by Hurtado [10]. He discussed simple beam with stochastic modulus of elasticity and deterministic load. Homogenous chaos expansion Radial basis (RBF) neural network was trained with some pairs of input and output by MCS simulations. Later, El-Beltagy et al. [11] developed this method to include the effect of random load and random modulus of elasticity using both of RBF neural network and polynomial chaos expansion (PCE).

Burgers' equation is an important partial differential equation from fluid dynamics, and is widely used for various physical applications, such as modeling of gas dynamics and traffic flow, shock waves [12], investigating the shallow water waves [13,14], in examining the chemical reaction diffusion model of Brusselator etc. [15]. In fact, it can be used as a model for any nonlinear wave propagation problem subject to dissipation [16]. Depending on the problem being modeled, this dissipation may result from viscosity, heat conduction, mass diffusion, thermal radiation, chemical reaction, or other source. Burgers' equation may be analyzed by using the exact shock-wave solution. In the current paper, finite-volume upwind technique is used to avoid the numerical instabilities and compute solution for small viscosity as discussed by Stephens et al. [17].

The SSFEM, and also the stochastic finite-volume, produces large sparse linear systems. Also, Burgers' equation with zero or small viscosity will be hyperbolic PDE which produces stiff linear system and requires very small time steps to be solved adequately. To enhance the performance of the developed solver, parallelization should be considered. Additionally, MCS should be of order 10⁴ or more for reliable comparisons. These simulations should be done in parallel as well. In the current paper, parallelization of both techniques is considered.

POLYNOMIAL CHAOS EXPANSION

Polynomial chaos expansion has many advantages in evaluating both statistical moments of any order and the p.d.f of system response which represents a complete solution of the random systems. Ghanem and Spanos [8], evaluated the system response as a summation of nonlinear functional of a set of $\{\xi_n(\theta)\}_{n=1}^{\infty}$ multiplied by deterministic constants. The system response in terms of polynomial chaos is written in the form:

$$\alpha(\theta) = a_0 \Gamma_0 + \sum_{i=1}^{\infty} a_i \Gamma_1(\xi_i) + \sum_{i=1}^{\infty}\sum_{j=1}^{\infty} a_{ij} \Gamma_2(\xi_i, \xi_j) \qquad (1)$$

$$+ \sum_{i=1}^{\infty}\sum_{j=1}^{\infty}\sum_{k=1}^{\infty} a_{ijk} \Gamma_3(\xi_i, \xi_j, \xi_k) + \cdots$$

Where Γ_p is the polynomial chaos of order p in a set of n random variables $\xi_i, \xi_j, \xi_k \ldots$ If the polynomial chaos are truncated at order p, the response of the system will be reduced to:

$$\alpha(\theta) = \sum_{i=0}^{p} c_i \Psi_i \left[\{\xi_n\}\right] \qquad (2)$$

where c_i is a set of deterministic coefficients, and $\psi_i[\{\xi_n\}]$ is a set of polynomials of random variables, these polynomials are orthogonal.

STOCHASTIC FINITE-VOLUME FORMULATION FOR BURGERS' EQUATION

Consider the one dimensional Burgers' equation [13,14]

$$\frac{\partial u}{\partial t} + u \frac{\partial u}{\partial x} = \upsilon \frac{\partial^2 u}{\partial x^2} \tag{3}$$

Subject to the following deterministic initial and boundary conditions:

$$u(x,0) = \phi(x); \ 0 \leq x \leq L$$
$$u(0,t) = \alpha; \ u(L,t) = \beta, \ t \geq 0 \tag{4}$$

When the time derivative term is dropped from equation, and for large viscosity, we are left with an elliptic partial differential equation representing the steady-state balance between the convective and diffusive terms. The difficulty in computing solutions to the Burgers' equation lies in the inability to effectively balance the nonlinear convective term, and the diffusive term. For zero (or small) viscosity, the equation tends to be hyperbolic and the solution technique should be adapted to account for the characteristics of the equation.

Considering v is a stochastic viscosity, the response will be also stochastic and they can both expanded using polynomial chaos expansion as:

$$\upsilon = \sum_{n=0}^{pc} \upsilon_n \Psi_n, \tag{5}$$

$$u = \sum_{i=0}^{pc} d_i \Psi_i$$

The number of polynomials (pc) is a function of the required order (p) and dimension (M). Then, the differential equation can be written as:

Upwind Finite-Volume Solution Of Stochastic Burgers' Equation

$$\sum_{i=0}^{pc}\left[\Psi_i \frac{\partial}{\partial t}d_i\right] + \sum_{i=0}^{pc}\left[\sum_{j=0}^{pc}\Psi_i \Psi_j d_j \frac{\partial}{\partial x}d_i\right] \quad (6)$$

$$= \sum_{i=0}^{pc}\left[\sum_{j=0}^{pc} v_j \Psi_i \Psi_j \frac{\partial^2}{\partial x^2}d_i\right]$$

Galerkin projection scheme can be applied to Equation (6) through multiplying both sides by Ψ_i and applying the expectation operator, yields:

$$\sum_{i=0}^{pc}\sigma_{\Psi_i}^2 \frac{\partial}{\partial t}d_i + \sum_{i=0}^{pc}\left[\sum_{j=0}^{pc}c_{ijl}d_j\frac{\partial}{\partial x}d_i\right] \quad (7)$$

$$= \sum_{i=0}^{pc}\left[\sum_{j=0}^{pc}c_{ijl}v_j\frac{\partial^2}{\partial x^2}d_i\right]$$

where, where, $\sigma_{\Psi_i}^2 = \langle \Psi_i \Psi_1 \rangle$, $c_{ii1} = \langle \Psi_i \Psi_i \Psi_1 \rangle$, then

$$\sum_{i=0}^{pc}\sigma_{\Psi_i}^2 \frac{\partial}{\partial t}d_i + \sum_{i=0}^{pc}\left[\sum_{j=0}^{pc}c_{ijl}\left(d_j\frac{\partial}{\partial x}d_i - v_j\frac{\partial^2}{\partial x^2}d_i\right)\right] = 0 \quad (8)$$

Using the finite-volume node-centered approach by integrating over the control volume, yields:

$$\sum_{i=0}^{pc}\sigma_{\Psi_i}^2 \frac{\partial}{\partial t}\int_{x_k - \frac{\Delta x}{2}}^{x_k + \frac{\Delta x}{2}} d_i \quad (9)$$

$$+ \sum_{i=0}^{pc}\left[\sum_{j=0}^{pc}c_{ijl}\left(\int_{x_k-\frac{\Delta x}{2}}^{x_k+\frac{\Delta x}{2}} d_j \frac{\partial}{\partial x}d_i - v_j \int_{x_k-\frac{\Delta x}{2}}^{x_k+\frac{\Delta x}{2}} \frac{\partial^2}{\partial x^2}d_i\right)\right] = 0$$

By using the upwind scheme for the face-centered values, then the differential equation is reduced to:

$$\sum_{i=0}^{pc}\left[\left(\sum_{j=0}^{pc}c_{ijl}\left(-d_j^n(k)-\frac{v_j}{\Delta x}\right)\right)d_i^{n+1}(k-1)\right.$$

$$+\left(\sigma_{\Psi_i}^2\frac{\Delta x}{\Delta t}+\sum_{j=0}^{pc}c_{ijl}\left(d_j^n(k)+\frac{2v_j}{\Delta x}\right)\right)d_i^{n+1}(k) \qquad (10a)$$

$$\left.+\left(\sum_{j=0}^{pc}c_{ijl}\left(-\frac{v_j}{\Delta x}\right)\right)d_i^{n+1}(k+1)\right]$$

$$=\sum_{i=0}^{pc}\sigma_{\Psi_i}^2\frac{\Delta x}{\Delta t}d_i^n(k) \quad \text{if} \quad d_i^n(k)\geq 0$$

or

$$\sum_{i=0}^{pc}\left[\left(\sum_{j=0}^{pc}c_{ijl}\left(-\frac{v_j}{\Delta x}\right)\right)d_i^{n+1}(k-1)\right.$$

$$+\left(\sigma_{\Psi_i}^2\frac{\Delta x}{\Delta t}+\sum_{j=0}^{pc}c_{ijl}\left(-d_j^n(k)+\frac{2v_j}{\Delta x}\right)\right)d_i^{n+1}(k) \qquad (10b)$$

$$\left.+\left(\sum_{j=0}^{pc}c_{ijl}\left(d_j^n(k)-\frac{v_j}{\Delta x}\right)\right)d_i^{n+1}(k+1)\right]$$

$$=\sum_{i=0}^{pc}\sigma_{\Psi_i}^2\frac{\Delta x}{\Delta t}d_i^n(k) \quad \text{if} \quad d_i^n(k)<0$$

Equations (10a) and (10b) result in sparse linear system. The linear system becomes larger as the order and/or the dimension of the polynomial chaos are increased. **Figure 1** shows the sparsity pattern when discretizing the domain into 100 divisions and for different values of the order and dimension. The number of nonzeros (NNZ) is shown below the figure of each case. Sparse storage should be considered to save the memory and to enhance the performance by using a suitable sparse linear solver.

SOLUTION TECHNIQUES AND DISCUSSION

Consider the one dimensional viscous Burgers' equation on the interval [0, 1] with Dirichlet boundary conditions; u = 1 at x = 0 and u = −1 at x = 1 respectively. The interval is discretized into N = 128 finite elements. Different values of order p and dimension M are used with different values of the mean viscosity v_0. In the current work, the stochastic variation of the viscosity is taken as:

$$v = \sum_{n=0}^{pc}(0.2v_0)^n \Psi_n$$

This means that the first stochastic component of the viscosity is only 20% of the mean value to avoid negative values of viscosity. Using the above proposed solver, we can notice that the effect of the mean viscosity on the solution. The larger the viscosity is, the smoother the solution is (parabolic behavior of Burgers' equation). On the other hand, as the mean viscosity decreases (and may reach zero) the response will be similar to the hyperbolic wave equation and a shock wave (with zero velocity) will be constructed. Figure 2 shows the mean solution using both the stochastic solver and the MCS simulations. The mean solutions are in a good agreement for different values of the mean viscosity.

Figure 3 shows the convergence history of the stochastic finite volume solver for different mean values of viscosity. The residual logarithm is reduced up to −10. The figure shows that the number of iterations required for convergence increases as the mean value of the viscosity is decreased. Consequently the CPU time increases for small values of viscosity due to the slow rate

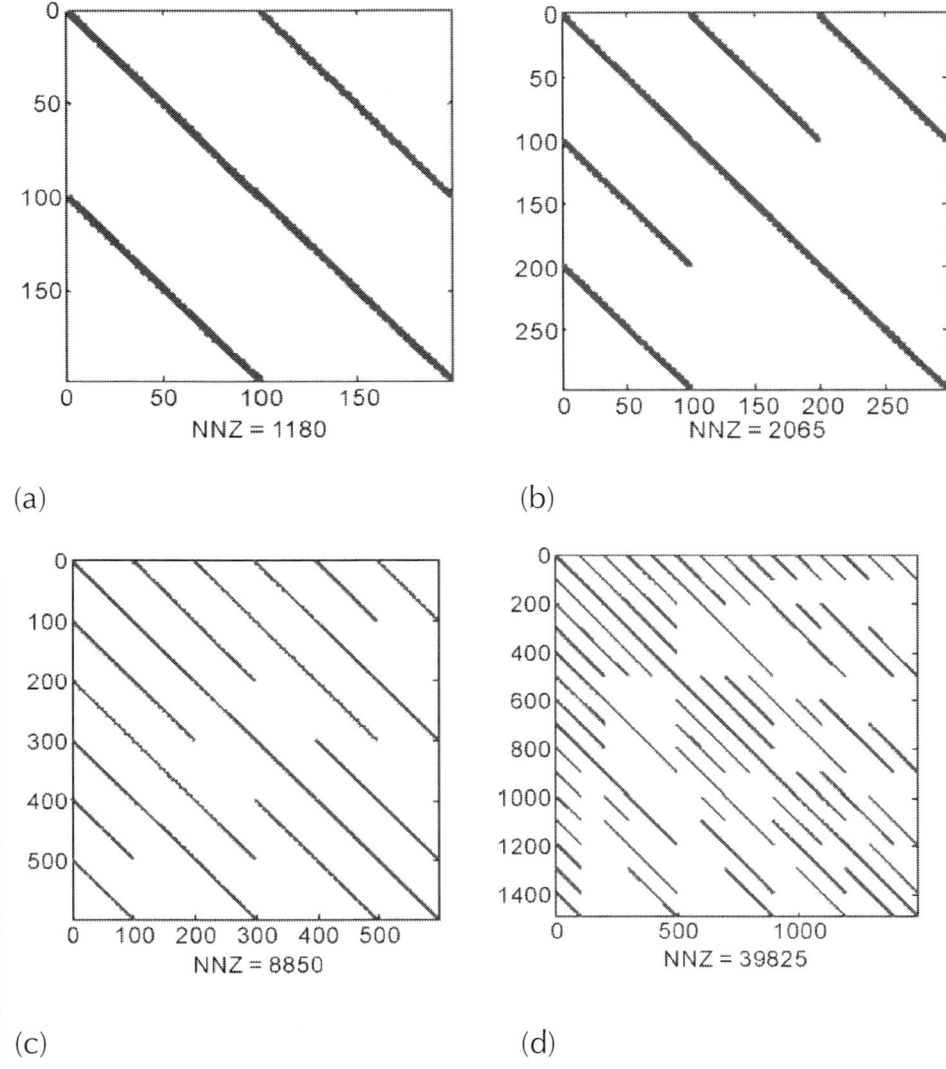

(a) (b) (c) (d)

Figure 1: Sparsity pattern for for different order (p) and different dimension (M). The number of nonzeros (NNZ) is shown. (a) p = 1, M = 1; (b) p = 1, M = 2; (c) p = 2, M = 2; (d) p = 2, M = 4.

Upwind Finite-Volume Solution Of Stochastic Burgers' Equation

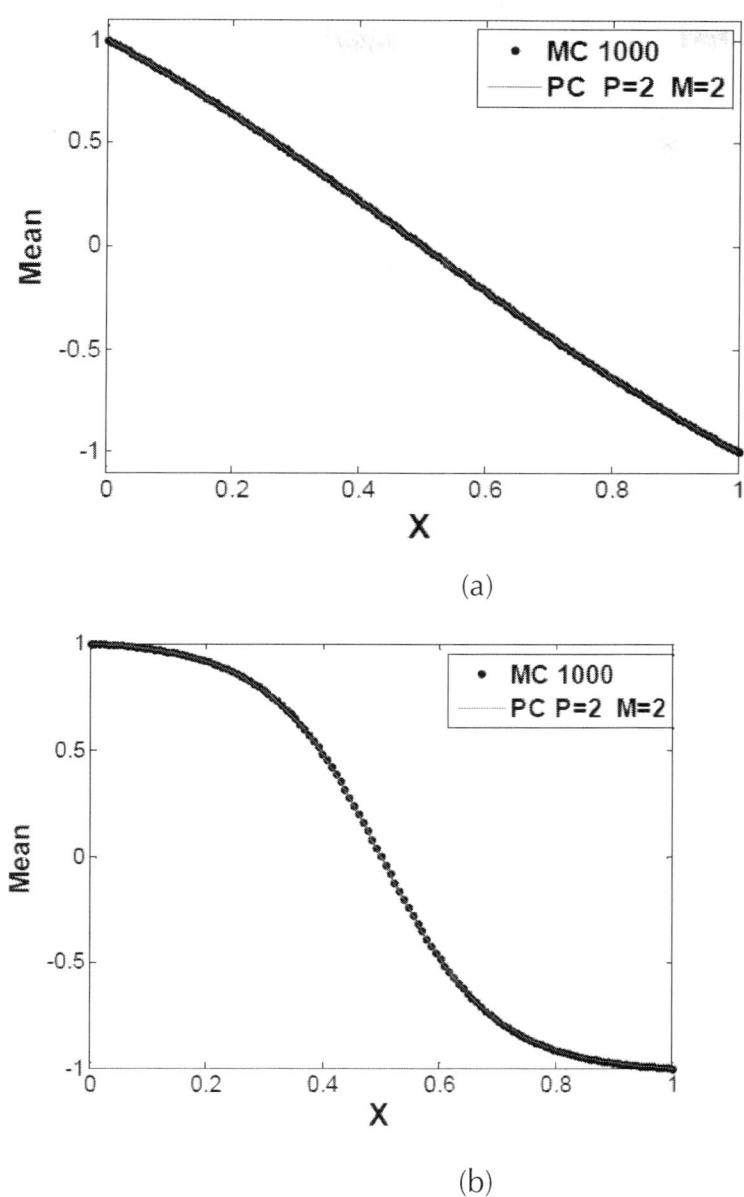

(a)

(b)

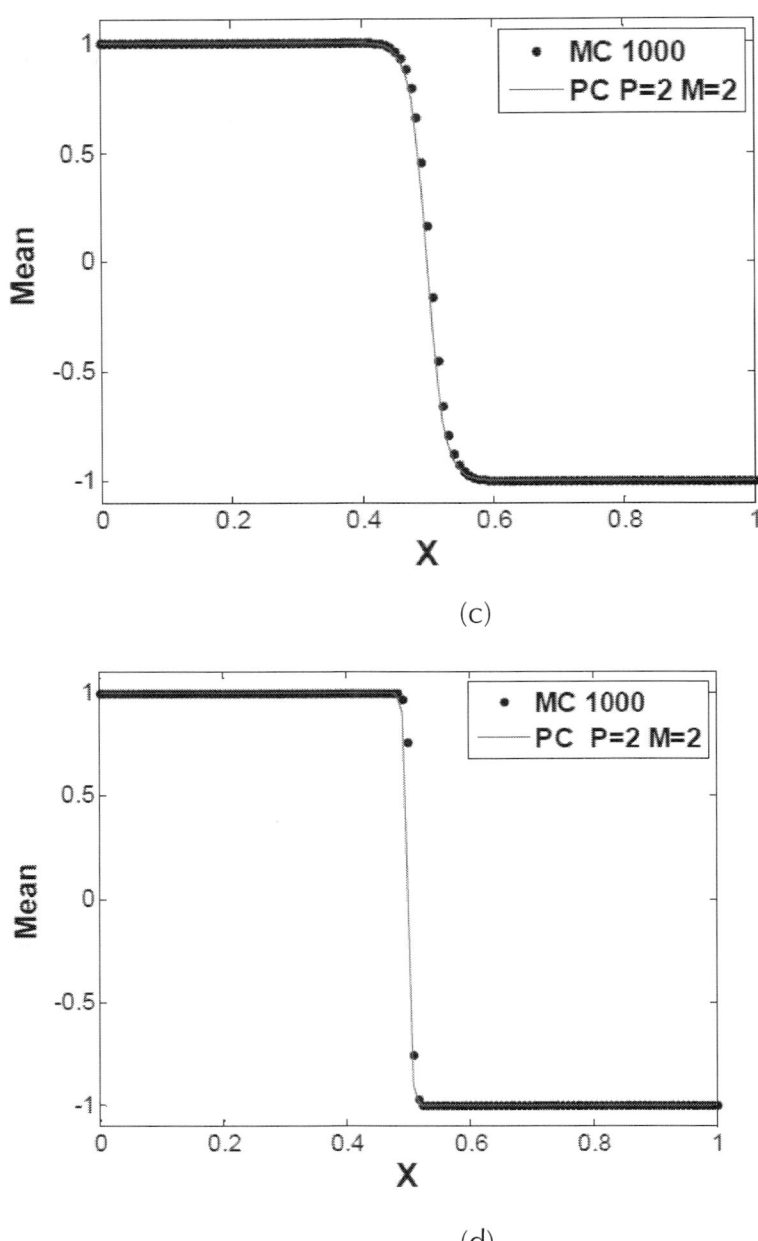

(c)

(d)

Figure 2: Mean value of the stochastic solver and MCS for different values of mean viscosity. (a) Viscosity = 1.0; (b) Viscosity = 0.1; (c) Viscosity = 0.01; (d) Viscosity = 0.001.

Upwind Finite-Volume Solution Of Stochastic Burgers' Equation

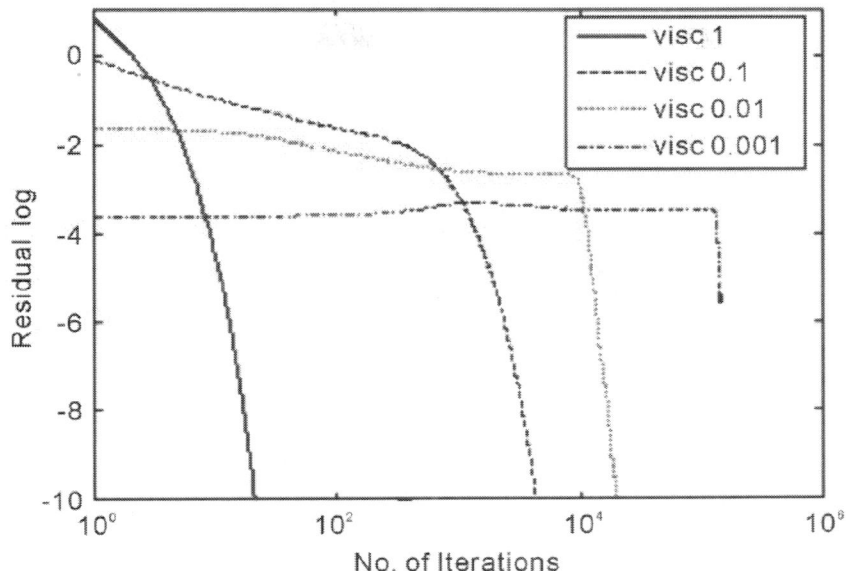

Figure 3: Solver convergence (Log-Log scale) for different values of mean viscosity with p = 2 and M = 2.

of convergence and smaller time steps. The CPU time needed to solve the stochastic system is listed in Table (1). The used workstation was intel® xean® CPU X5690 3.47 GHz (6 cores), 8 GB RAM, 64-bit O.S. The PARDISO (Parallel Direct Solver) [18,19] is used in solving the stochastic linear system. The parameters of the PARDISO solver are set to utilize the 6 cores available on the workstation. More than 50% performance increase is obtained in solving the linear system when using PARDISO.

Additionally, the stochastic linear system can be constructed in parallel, but this was not done in the current work as it will consume more memory storage. The MCS simulations can be done by generating random normal distribution for a certain mean viscosity and use these values to run the deterministic finite-volume solver. The deterministic solver can be developed in a similar way as the above described stochastic solver. In the current work, the stochastic solver with zero order and zero dimension is used instead. This will assure that all parameters are the same when comparing the two solution techniques (Stochastic and MCS). The MCS simulations are done also in parallel. The MCS simulations are independent runs and hence the parallelization is straight forward using Open-MP support available with the current C++ compilers. Histograms of the stochastic values of viscosity with mean value of 0.1 is shown in **Figure 4.**

Table 1 shows the CPU time comparisons between the stochastic solver and the MCS simulations. It can be notice that using of stochastic solver decrease CPU time dramatically with acceptable accuracy.

The standard deviation of stochastic response of Burgers' equation for different values of viscosity compared with MCS is shown in Figures 5 and 6. It can be noticed that the standard deviation shrinks vanishes around the midpoint for all values of the mean viscosity. The MCS simulations are in a good agreement with the stochastic solver for larger mean values of the viscosity. As the

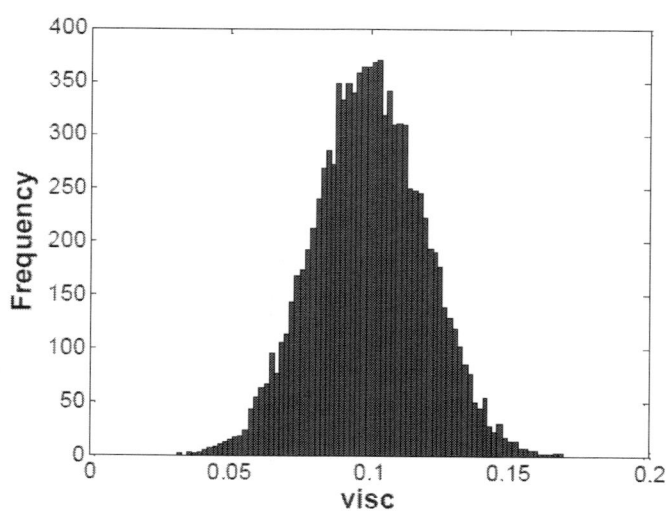

Figure 4: Histogram of the random values of viscosity used in MCS with mean value = 0.1.

Upwind Finite-Volume Solution Of Stochastic Burgers' Equation

Table 1: Time comparison (seconds) between MCS and the stochastic solver for different mean viscosity, different or-der (p) and dimension (M).

	visc = 0.1	visc = 0.01
MCS: 100	29.33	3.69
MCS: 1000	233.48	30.9
MCS: 10000	1991.33	303.56
PCE: $p = 1$ $M = 1$	2.9	1.25
PCE: $p = 2$ $M = 2$	11.8	7.27

mean viscosity decreases, the deviations with the MCS are increased. **Figure 5**(d) is the same as **Figure 5**(c) but with different domain to show the deviations when using both solution techniques. Similarly, Figures 6(a) and (b) are the same but with different domains.

Figure 7 shows the relative error in the mean and the standard deviation when using the two techniques. As it is shown in the figure, the relative error in the mean increases as the mean viscosity decreases. On the other hand, the relative error in the standard deviation decreases as the mean viscosity decreases.

Figure 8 shows the first stochastic solution component of the polynomial chaos expansion (d_1) for different values of the mean viscosity. As the viscosity decreases, the first stochastic component becomes localized around the midpoint and its magnitude increases. The other (higher) stochastic components are small compared to the first component. So, the first component is approximately equals to the standard deviation.

Figure 9 shows the stochastic solution (mean plus first stochastic component) for different values of the mean viscosity. The first component is scaled up by a factor of 10 to clarify the effect on the stochastic response. There are irregularities in the solution due to the random variation of the viscosity. These irregularities may become very sharp even around the shock wave.

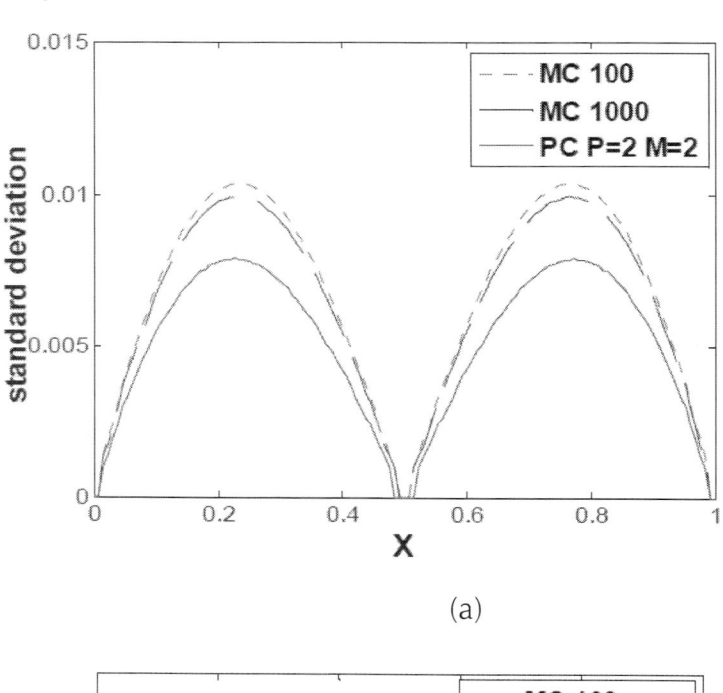

(a)

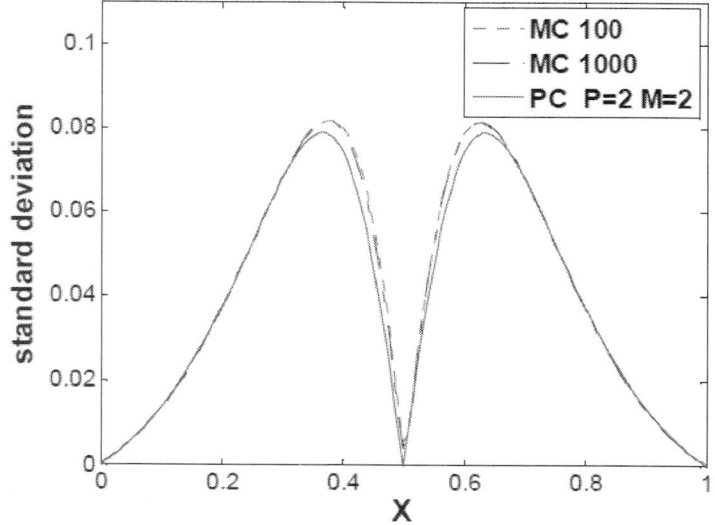

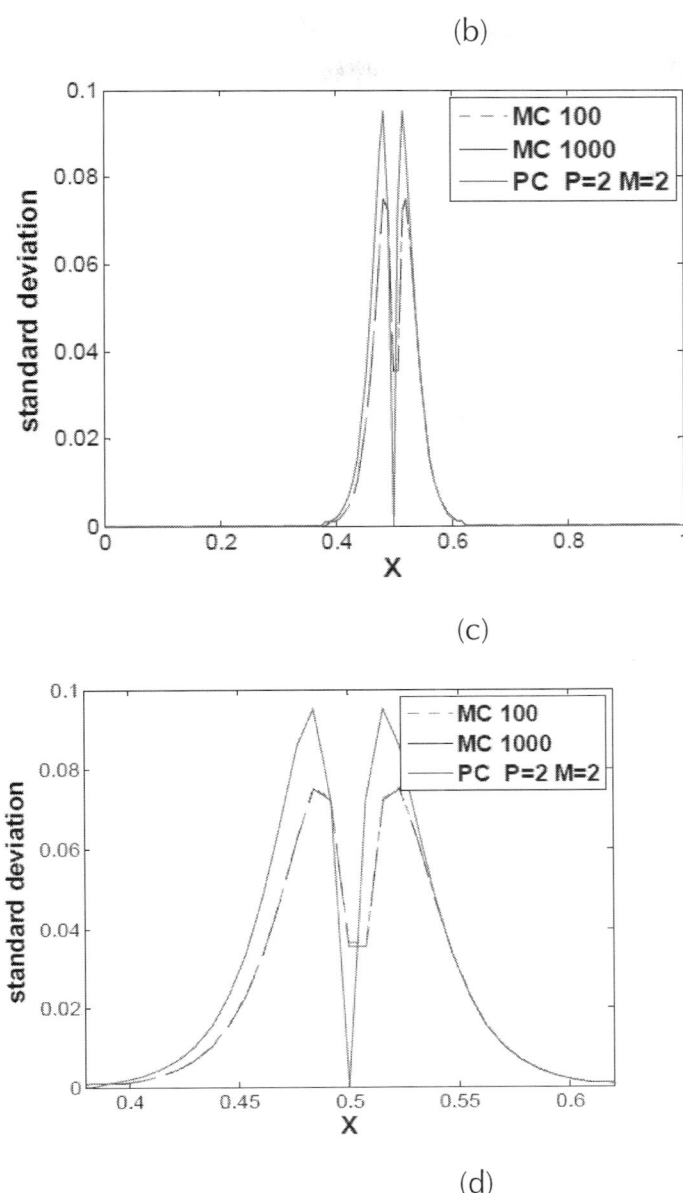

Figure 5: Standard deviation of the stochastic solver and MCS for different values of mean viscosity. (a) viscosity = 1.0; (b) viscosity = 0.1; (c) viscosity = 0.01; (d) viscosity = 0.01.

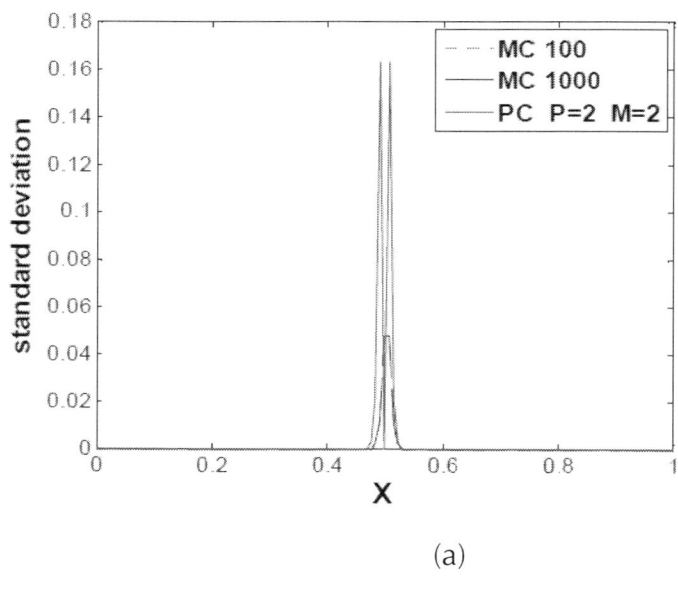

(a)

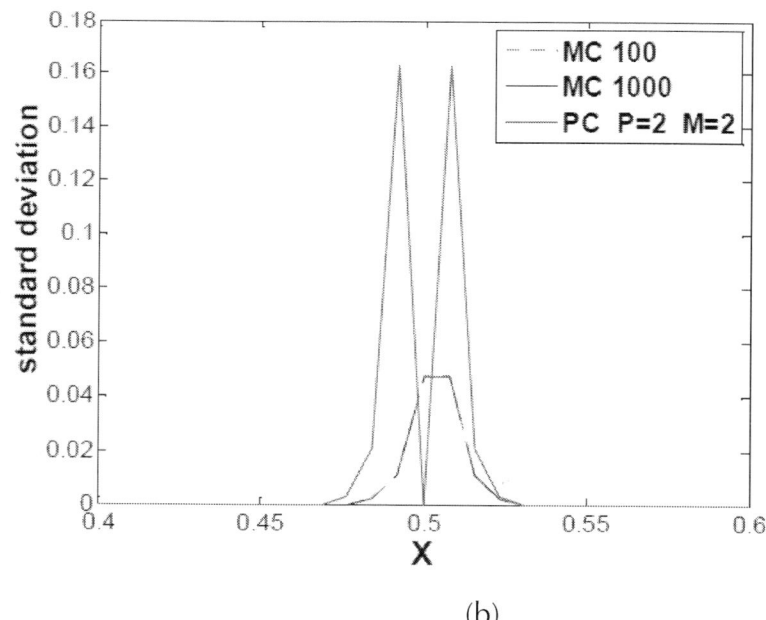

(b)

Figure 6: Standard deviation of the stochastic solver and MCS for mean viscosity = 0.001. (a) and (b) are the same but different domain.

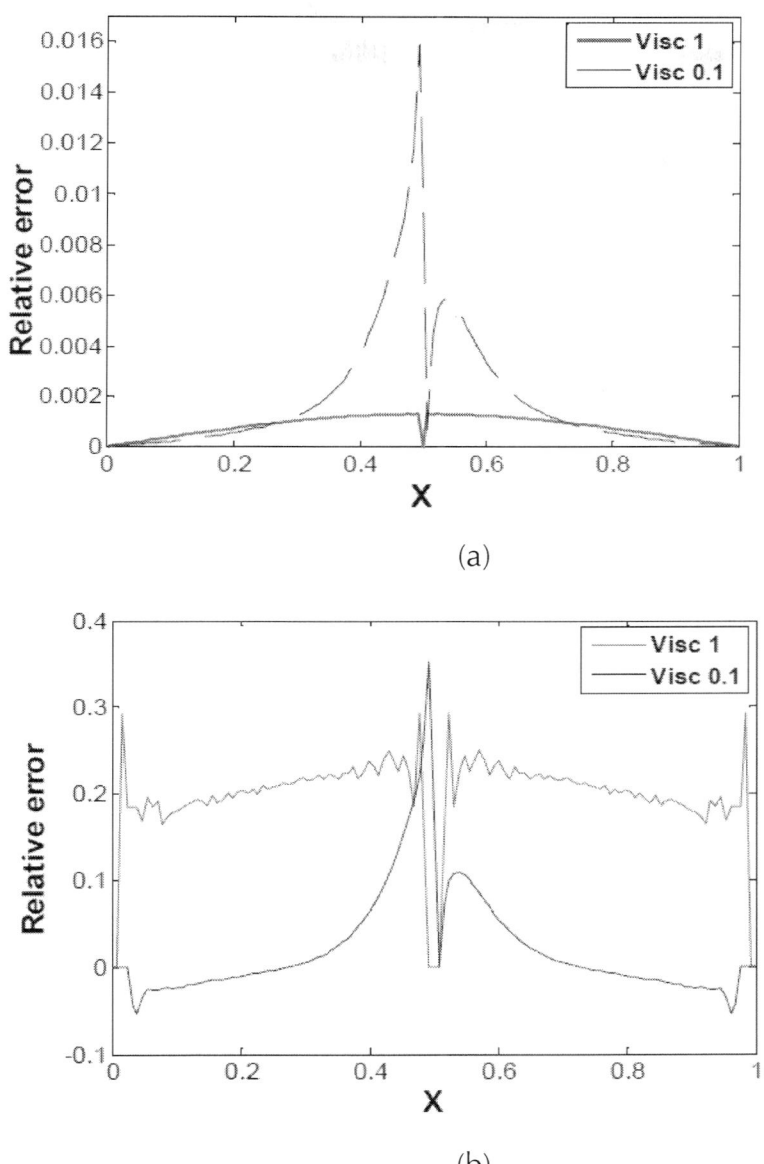

Figure 7: Relative error for the mean (a) and the standard deviation (b) using the stochastic solver and MCS for dif-ferent values of viscosity.

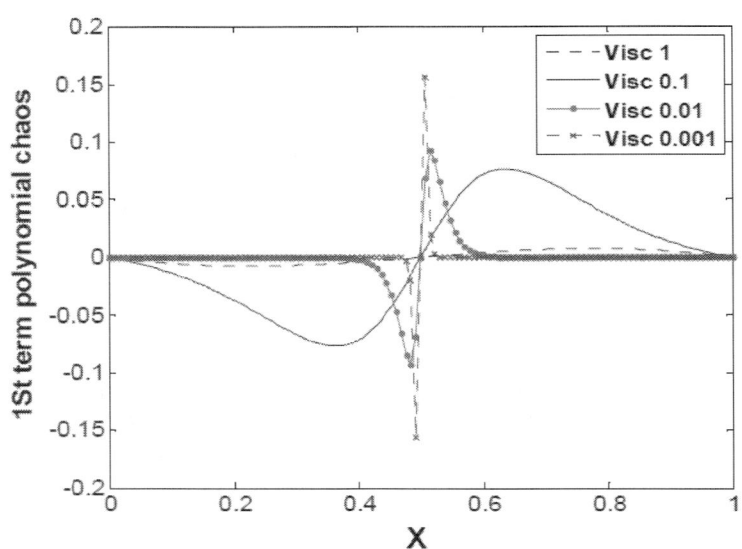

Figure 8: The first term polynomial chaos d1 of the stochas-tic response for different values of the mean viscosity with p = 2 and M = 2.

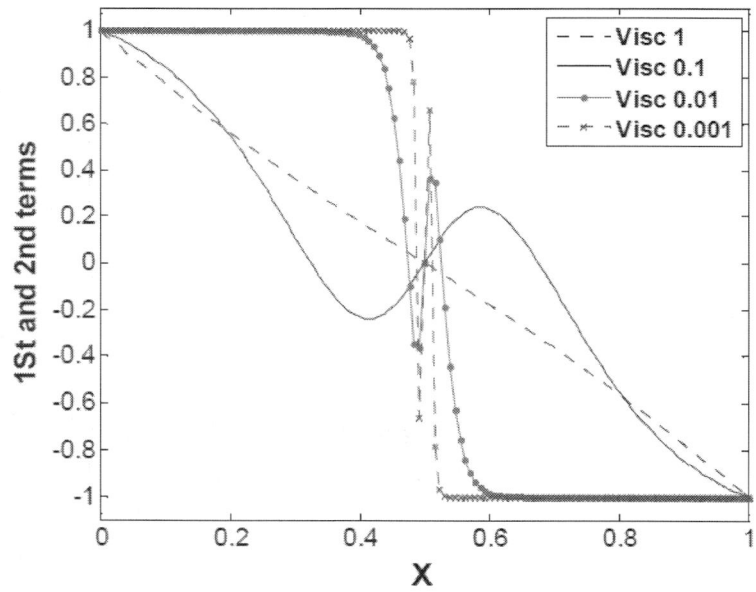

Figure 9: Mean solution plus the first stochastic component of the response obtained for different values of viscosity (the 1st stochastic component is scaled by a factor of 10).

Figure 10 illustrates the p.d.f of the solution at selected nodes for different values of the mean viscosity. The p.d.f is in a good agreement for larger values of the mean viscosity. As the mean viscosity decreases, the p.d.fs from both techniques deviate. The minimum and the maximum values of the response can be obtained easily, this issue is very important in the design stage and for reliability and safely analysis.

CONCLUSION

The stochastic finite-volume solution has advantages in evaluating the p.d.f of the system response with minimum cost. The developed solver based on the polynomial chaos expansion succeeds in analyzing stochastic nonlinear systems with high performance. Using the upwind scheme was proven as an appropriate choice to handle the system in the parabolic regime and also in the hyperbolic regime. The MCS simulations deviate from the stochastic solution when the system tends to be hyperbolic (mean viscosity decreases). Parallelization of the stochastic solver is important to increase the performance especially when solving the resulting linear sparse system. The stochastic solver developed in this work can be extended to higher dimensions in a straight forward way.

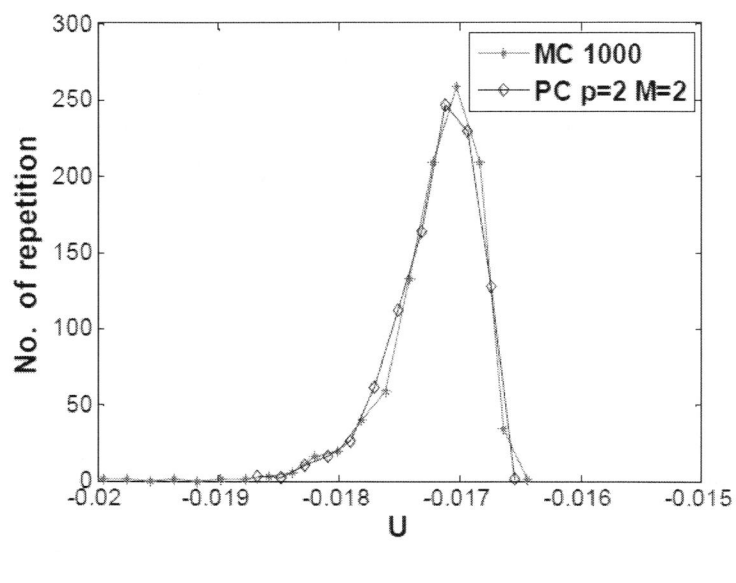

(a)

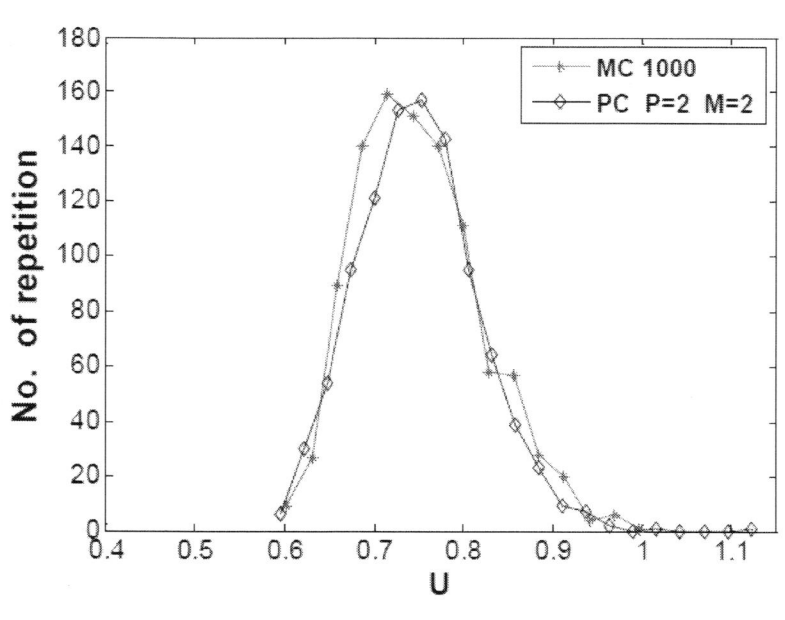

(b)

Upwind Finite-Volume Solution Of Stochastic Burgers' Equation

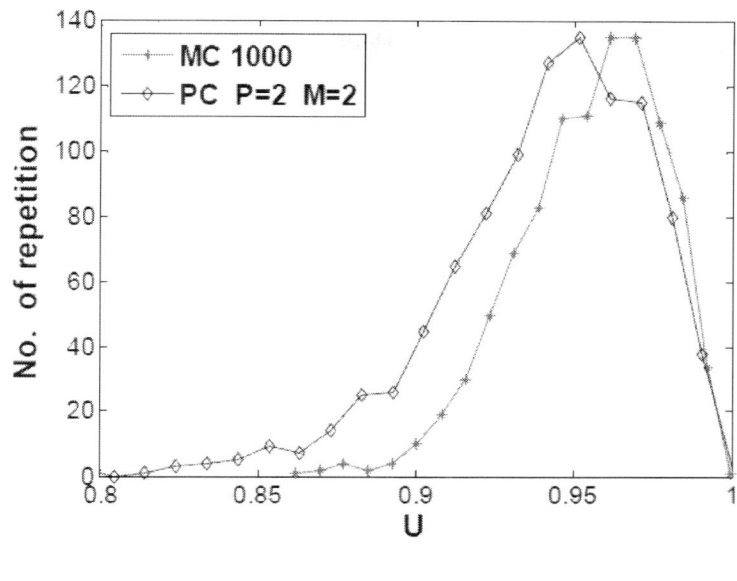

(c)

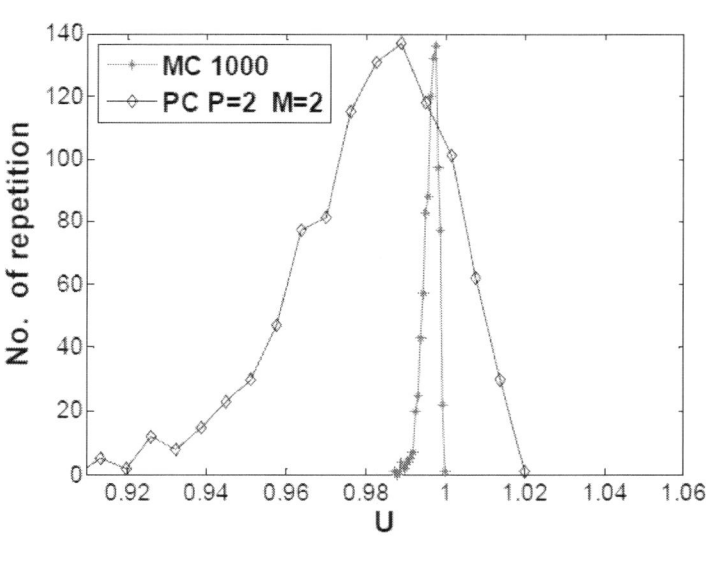

(d)

ACKNOWLEDGMENTS

We should thank Prof. Magdy El-Tawil for his supervision and guidance to this work.

REFERENCES

1. O. H. Galal, W. El-Tahan, M. A. El-Tawil and A. A. Mahmoud, "Spectral SFEM Analysis of Structures with Stochastic Parameters under Stochastic Excitation," Structural Engineering and Mechanics, Vol. 28 No. 3, 2008, pp. 281-294.
2. O. H. Galal, "The Solution of Stochastic Linear Partial Differential Equation Using SFEM through Neumann and Homogeneous Chaos Expansions," Ph.D. Thesis, Cairo University, Cairo, 2005.
3. S. Rahman and H. Xu, "A Meshless Method for Computational Structure Mechanics," International Journal for Computational Methods in Engineering Science and Mechanics, Vol. 6, No. 1, 2005, pp. 41-58. doi:10.1080/15502280590888649
4. M. Kaminski, "Stochastic Perturbation Approach to Engineering Structure Variability by the Finite Difference Method," Journal of Sound and Vibration, Vol. 251 No. 4, 2002, pp. 651-670. doi:10.1006/jsvi.2001.3850
5. G. Stefanou, "The Stochastic Finite Element Methods: Past, Present and Future," Computer Methods in Applied Mechanics and Engineering, Vol. 198, No. 9-12, 2009, pp. 1031-1051. doi:10.1016/j.cma.2008.11.007
6. M. Shinozuka and T. Nomoto, "Response Variability Due to Spatial Randomness of Material Properties," Technical Report, Columbia University, New York, 1980.
7. A. Henriques, J. Veiga, J. Matos and J. Delgado, "Uncertainty Analysis of Structural Systems by Perturbation Techniques," Journal of Structural and Multidisciplinary Optimization, Vol. 35 No. 3, 2008, pp. 201-212. doi:10.1007/s00158-007-0218-z
8. R. Ghanem and P. D. Spanos, "Stochastic Finite Elements: A Spectral Approach," 2nd Edition, Dover, New-York, 2002.
9. H. Panayirci, "Computational Strategies for Efficient Stochastic Finite Element Analysis of Engineering Structures," Ph.D. Thesis, University of Innsbruck, Innsbruck, 2010.
10. J. Hurtado, "Analysis of One Dimensional Stochastic Finite Element Using Neural Network," Probabilistic Engineering Mechanics, Vol.17, No. 1, 2001, pp. 35-44.
11. M. A. El-Beltagy, O. H. Galal and M. I. Wafa, "Uncertainty Quantification of a 1-D Beam Deflection Due to Stochastic Parameters," International Conference on Numerical Analysis and Applied Mathematics, Halkidiki, 19-25 September 2011, pp. 2000-2003.doi:10.1063/1.3637007.
12. J. D. Cole, "On a Quasilinear Parabolic Equations Occurring in Aerodynamics," Quarterly of Applied Mathematics, Vol. 9, 1951, pp. 225-236.

13. J. D. Logan, "An Introduction to Nonlinear Partial Differential Equations," Wily-Interscience, New York, 1994.
14. L. Debtnath, "Nonlinear Partial Differential Equations for Scientist and Engineers," Birkhauser, Boston, 1997.
15. G. Adomian, "The Diffusion-Brusselator Equation," Computers & Mathematics with Applications, Vol. 29, No. 5, 1995, pp. 1-3. doi:10.1016/0898-1221(94)00244-F
16. C. Fletcher, "Burgers' Equation: A Model for All Reasons," Numerical Solutions of Partial Differential Equations, North-Holland Pub. Co., Holland, 1982.
17. A. B. Stephens, R. B. Kellogg and G. R. Shubin, "Uniqueness and the Cell Reynolds Number," SIAM Journal on Numerical Analysis, Vol. 17, No. 6, 1980.
18. O. Schenk and K. Gärtner, "Solving Unsymmetric Sparse Systems of Linear Equations with PARDISO," Journal of Future Generation Computer Systems, Vol. 20, No. 3, 2004, pp. 475-487. doi:10.1016/j.future.2003.07.011
19. O. Schenk and K. Gärtner, "On Fast Factorization Pivoting Methods for Symmetric Indefinite Systems," Electronic Transactions on Numerical Analysis, Vol. 23, 2006, pp. 158-179.

CITATION

Mohamed A. El-Beltagy, Mohamed I. Wafa, Osama H. Galal Upwind Finite-Volume Solution of Stochastic Burgers' Equation DOI:10.4236/am.2012.331247

Solution of Nonlinear Stochastic Langevin's Equation Using WHEP, Pickard and HPM Methods

Maha Hamed[1], Magdy A. El-Twail[2], Beih El-desouky[1], and Mohamed A. El-Beltagy[3]

[1]Department of Mathematics, Faculty of Science & Mansoura University, Mansoura, Egypt
[2]Department of Engineering Mathematics & Physics, Engineering Faculty, Cairo University, Giza, Egypt
[3]Department of Electrical & Computer Engineering, Engineering Faculty, Effat University, Jeddah, KSA

ABSTRACT

This paper introduces analytical and numerical solutions of the nonlinear Langevin's equation under square nonlinearity with stochastic non-homogeneity. The solution is obtained by using the Wiener-Hermite expansion with perturbation (WHEP) technique, and the results are compared with those of Picard iterations and the homotopy perturbation method (HPM). The WHEP technique is used to obtain up to fourth order approximation for different number of corrections. The mean and variance of the solution are obtained and compared among the different methods, and some parametric studies are done by using Matlab.

INTRODUCTION

Stochastic differential equations based on the white noise process provide a powerful tool for dynamically modeling complex and uncertain aspects. Developing efficient numerical methods for simulating SPDEs is a very important while challenging research topic. The study of random solutions of partial differential equations was developing continuously since Kampe de Feriet in 1955 [1] and Bharucha-Reid in 1965 [2]. It was 1973 when Lo Dato V. had discussed the mathematical problems associated with Navier-Stokes equation [3], a general solution for the heat conducting problem with random source, initial and boundary conditions that was introduced by B. A. Georges [4]. The Langevin's equation [5] with square nonlinear losses and stochastic non-homogeneity is solved by using different techniques, mainly the WHEP technique [6-11], the Picard approximation [8-10] and HPM [9-11].

It is appropriate to assume that the nonlinear term in SPDE affecting the phenomena under study is small enough; then its intensity is controlled by means of a frank small parameter, saying ε [12]. In (WHE) approach, there is no randomness directly involved in the computations.

The application of WHE aims to find a truncated series of the solution process of the differential equation. The truncated series composes of two major parts: the first is the Gaussian part which consists of the first two terms, while the rest of the series constitutes the non-Gaussian part. In nonlinear cases, there exist difficulties in solving the resultant set of the deterministic set of integro-differential obtained after the direct application of WHE. The WHEP technique was introduced in [6] using perturbation method to solve perturbed nonlinear equation.

The WHE was developed by Norbert Wiener [13]. Wiener constructed an orthonormal random base for expanding homogeneous chaos depending on white noise and used it to study problems in statistical mechanics. Cameron and Martin [14] developed a more explicit and intuitive formulation for Wiener-Hermite expansion, and this technique is now known as Wiener-Chaos expansion (WCE). Their development was based on an explicit discretization of the white noise process through its Fourier expansion, which was missed in Wiener's original formalism. Since Cameron and Martin's work, WHE has become a useful tool in stochastic analysis involving white noise (Brownian motion) [14]. Also, another formulation was suggested and applied by Meecham and his co-workers [15,16]. They have developed a theory of turbulence involving a truncated WHE of the velocity field. This technique was also used by M. El-Tawil and his co-workers [6-10].

The main goal of this paper is to apply the WHEP technique in solving the nonlinear stochastic Langevin's equation and compare the solution with those of Picard's and HPM techniques. In Section 2, the problem will be introduced; in Section 3, the application of WHEP technique is considered; the Picard approximation technique is applied in Section 4; Section 5 describes the application of HPM method; Section 6 represents the results for every method; and Section 7 shows comparisons between the different techniques.

PROBLEM FORMULATION

In this paper, the nonlinear stochastic Langevin's equation with external stochastic excitation in the form of white noise is considered. It formulates general relation between the velocity $\upsilon(t;\omega)$ of a unit mass particle and the non homogeneity random forces $F(t;\omega)$ of a fluid medium with phenomenological coefficient $\varepsilon(\omega)$ such that [8,17]:

$$\frac{d\upsilon}{dt}+\varepsilon(\omega)\upsilon^q = F(t;\omega), \ \upsilon(0;\omega)=v_0, \ t\geq 0. \tag{1}$$

The variable ω is a random output of a triple probability space (Ω,B,P), where Ω is a sample space, B is a σ-algebra associated with Ω and P is a probability measure. For simplicity, ω will be dropped later on.

The following assumptions are taken for simplicity:

- Constant phenomenological coefficient; $\varepsilon(\omega) = \varepsilon$.
- deterministic initial condition; $\upsilon(0,\omega) = v_0$.
- Nonlinearity of second order; q=2.
- Medium forces in terms of white noise n(t; ω)=n(t) enveloped by a deterministic function f(t) and scaled by scalar amplitude ρ i.e. F(t; ω)=ρ f(t)n(t; ω).

Under the above hypothesis; the model becomes

$$\frac{d\upsilon}{dt} + \varepsilon \upsilon^q = \rho f(t)n(t), \quad \upsilon(0;\omega) = v_0, \quad t \geq 0. \tag{2}$$

Theorem: The solution of Equation (2), if exists, is a power series in ε when using Picard approximation technique i.e.

$$\upsilon(t) = \sum_{i=0}^{\infty} \varepsilon^i \upsilon_i(t).$$

The theorem can be proved using the mathematical induction with the Picard iterative technique as in [7].

The deterministic solution of Equation (2) at [8], is:

$$\upsilon(t) = v_0 (1 + \varepsilon v_0 t)^{-1}, \tag{3}$$

which can be expanded as follows,

$$\upsilon(t) = v_0 \left(1 - \varepsilon v_0 t + \varepsilon^2 v_0^2 t^2 - \varepsilon^3 v_0^3 t^3 + \cdots \right), \quad |\varepsilon v_0 t| < 1. \tag{4}$$

The solution is a series of successive addition and subtractions of the ascendant orders of the quantity $\varepsilon v_0 t$ resulting in oscillatory behavior around the exact solution; the motion within a medium with highly fractional forces will vanish rapidly compared to a medium with lower fractional forces [8].

WHEP TECHNIQUE

Due to the completeness of the Wiener-Hermite set [15], any arbitrary stochastic process $\upsilon(t;\omega)$ can be expanded as:

$$\upsilon(t;\omega) = \upsilon^{(0)}(t) + \sum_{k=1}^{\infty} \int_{R^k} \upsilon^{(k)}(t;t_1,\cdots,t_k) H^{(k)}(t_1,\cdots,t_k) d\tau_k, \tag{5}$$

where $d\tau_k = dt_1 dt_2 \cdots dt_k$ and $\int$ is a k-dimensional integral over the disposable variables $t_1, t_2, \ldots, t_k$. The functional $H^{(n)}(t_1,t_2,\ldots,t_n)$ is the nth order Wiener-Hermite time-independent functional. The Wiener-Hermite functional form a complete set with $H^{(0)}=1$ and $H^{(1)}(t)=n(t)$ which is the white noise. The WHE method can be used in solving Equation (2). A set of deterministic integro-differential equations are obtained in the de- terministic kernels [7],

$$\upsilon^{(i)}(t;t_1,t_2,\ldots,t_i); i \geq 0$$

To obtain approximate solutions of these deterministic kernels, one can use perturbation technique in the case of having a perturbed system depending on a small parameter . Expanding the kernels as a power series of another set of simpler iterative equations in the kernel series components are obtained.

If $\upsilon^{(k)} = \sum \varepsilon^i \upsilon_i^{(k)}$ then it will be convergent if [7]:

$$|\varepsilon| \leq \left|\frac{\upsilon_i^{(k)}}{\upsilon_{i+1}^{(k)}}\right|, \text{ for } t \in [t_0, T].$$

where NC refers to the number of corrections, m is the order, the statistical properties of the solution can be calculated as [7]:

$$E[\upsilon(t)] = \sum_{i=0}^{NC} \varepsilon^i \upsilon_i^{(0)}, \text{Var}[\upsilon(t)] = \sum_{k=1}^{m}(k!) \int_{R^k} \left(\sum_{i=0}^{NC} \varepsilon^i \upsilon_i^{(k)}\right)^2 d\tau_k. \quad (6)$$

Applying the WHEP algorithm [18] to get the deterministic set of differential equations. The envelop function f(t) which will be chosen as a constant function f(t)=1 and as a decaying exponential function $f(t) = e^{-\lambda t}$. The initial conditions are assumed deterministic as follows for $t_1 \leq t$:

For NC = 1: we have the following

$$\dot{\upsilon}_0^{(0)}(t) = 0, \ \upsilon(0) = v_0, \qquad \upsilon_0^{(0)}(t) = v_0,$$

$$\dot{\upsilon}_0^{(1)}(t;t_1) = \rho f(t)\delta(t-t_1), \ \upsilon_0^{(1)}(t;t_1) = \rho f(t_1),$$

$$\dot{v}_0^{(2)}(t;t_1,t_2) = 0, \qquad v_0^{(2)}(t;t_1,t_2) = 0,$$

$$\dot{v}_0^{(3)}(t;t_1,t_2,t_3) = 0, \qquad v_0^{(3)}(t;t_1,t_2,t_3) = 0,$$

$$\dot{v}_0^{(4)}(t;t_1,t_2,t_3,t_4) = 0, \qquad v_0^{(4)}(t;t_1,t_2,t_3,t_4) = 0,$$

$$\dot{v}_1^{(0)}(t) + \left[\dot{v}_0^{(0)}(t)\right]^2 + \int_0^t \left[v_0^{(1)}(t;t_1)\right]^2 dt_1 = 0, \; v_1^{(0)}(t) = -v_0^2 t - \rho^2 \int_0^t \int_0^s f^2(t_1,s) dt_1 ds, \qquad (7)$$

$$\dot{v}_1^{(1)}(t;t_1) + 2v_0^{(0)}(t)v_0^{(1)}(t;t_1) = 0, \quad v_1^{(1)}(t;t_1) = -2v_0\rho t f(t_1),$$

$$\dot{v}_1^{(2)}(t;t_1,t_2) + v_0^{(1)}(t;t_1)v_0^{(1)}(t;t_2) = 0, \quad v_1^{(2)}(t;t_1,t_2) = -\rho^2 t f(t_1) f(t_2),$$

$$\dot{v}_1^{(3)}(t;t_1,t_2,t_3) = 0, \qquad v_1^{(3)}(t;t_1,t_2,t_3) = 0,$$

$$\dot{v}_1^{(4)}(t;t_1,t_2,t_3,t_4) = 0, \qquad v_1^{(4)}(t;t_1,t_2,t_3,t_4) = 0.$$

For NC = 2: We have the above equations in addition to:

$$\dot{v}_2^{(0)}(t) + 2v_0^{(0)}(t)v_1^{(0)}(t) + 2\int_0^t v_0^{(1)}(t;t_1)v_1^{(1)}(t;t_1) dt_1 = 0,$$

$$v_2^{(0)}(t) = v_0^3 t^2 + 2v_0\rho^2 \int_0^t \int_0^u \int_0^s f^2(t_1,s,u) dt_1 ds du + 4v_0\rho^2 \int_0^t s \int_0^s f^2(t_1,s) dt_1 ds,$$

$$\dot{v}_2^{(1)}(t;t_1) + 2v_0^{(0)}(t)v_1^{(1)}(t;t_1) + 2v_1^{(0)}(t)v_0^{(1)}(t;t_1) + 4\int_0^t v_0^{(1)}(t;t_2)v_1^{(2)}(t;t_1,t_2) dt_2 = 0,$$

$$v_2^{(1)}(t;t_1) = \left\{3v_0^2\rho t^2 + 4\rho^3 \int_0^t s \int_0^s f^2(t_1,s) dt_1 ds + 2\rho^3 \int_0^t \int_0^u \int_0^s f^2(t_1,s,u) dt_1 ds du\right\} f(t_1),$$

(8)

$$2\dot{v}_2^{(2)}(t;t_1,t_2) + 2v_0^{(1)}(t;t_1)v_1^{(1)}(t;t_2) + 2v_1^{(1)}(t;t_2)v_0^{(1)}(t;t_1) + 4v_1^{(0)}(t)v_1^{(2)}(t;t_1,t_2) = 0,$$

$$v_2^{(2)}(t;t_1,t_2) = 3v_0\rho^2 t^2 f(t_1) f(t_2), 6\dot{v}_2^{(3)}(t;t_1,t_2,t_3) + 12\left[v_0^{(1)}(t;t_1)v_1^{(2)}(t;t_2,t_3)\right] = 0,$$

$$v_2^{(3)}(t;t_1,t_2,t_3) = \rho^3 t^2 f(t_1) f(t_2) f(t_3), 24\dot{v}_2^{(4)}(t;t_1,t_2,t_3,t_4) = 0, \; v_2^{(4)}(t;t_1,t_2,t_3,t_4) = 0.$$

Solution of Nonlinear Stochastic Langevin's Equation Using

For NC = 3: We have the above equations in addition to:

$$\dot{v}_3^{(0)}(t) + \left[v_1^{(0)}(t)\right]^2 + 2v_0^{(0)}(t)v_2^{(0)}(t) + \int_0^t \left[v_1^{(1)}(t;t_1)\right]^2 dt_1$$

$$+ 2\int_0^t v_0^{(1)}(t;t_1)v_2^{(1)}(t;t_1)dt_1 + 2\int_0^t\int_0^t \left[v_1^{(2)}(t;t_1,t_2)\right]^2 dt_1 dt_2 = 0,$$

$$v_3^{(0)}(t) = -v_0^4 t^3 - \rho^4 \int_0^t\left[\int_0^u\int_0^s f^2(t_1,s)dt_1 ds\right]^2 du - 2v_0^2\rho^2 \int_0^t u \int_0^u\int_0^s f^2(t_1,s,u)dt_1 dsdu$$

$$- 4v_0^2\rho^2 \int_0^t\int_0^u\int_0^v\int_0^s f^2(t_1,s,v,u)dt_1 dsdvdu - 8v_0^2\rho^2 \int_0^t\int_0^u s\int_0^s f^2(t_1,s,u)dt_1 dsdu$$

$$- 10v_0^2\rho^2 \int_0^t s^2 \int_0^s f^2(t_1,s)dt_1 ds - 4\rho^4 \int_0^t\left[\int_0^u\int_0^v\int_0^s f^2(t_1,s,v)dt_1 dsdv \int_0^u f^2(t_1,u)dt_1\right]du$$

$$- 8\rho^4 \int_0^t\left[\int_0^u s\int_0^s f^2(t_1,u)dt_1 ds \int_0^u f^2(t_1,u)dt_1\right]du - 2\rho^4 \int_0^t s^2 \int_0^s\int_0^s f^2(t_1,s)f^2(t_2,s)dt_1 dt_2 ds,$$

$$\dot{v}_3^{(1)}(t;t_1) + 2v_0^{(0)}(t)v_2^{(1)}(t;t_1) + 2v_1^{(0)}(t)v_1^{(1)}(t;t_1) + 2v_2^{(0)}(t)v_0^{(1)}(t;t_1)$$

$$+ 4\int_0^t v_0^{(1)}(t;t_2)v_2^{(2)}(t;t_1,t_2)dt_2 + 4\int_0^t v_1^{(1)}(t,t_2)v_1^{(2)}(t,t_1,t_2)dt_2 = 0,$$

$$v_3^{(1)}(t;t_1) = \left[-4v_0^3\rho t^3 - 8v_0\rho^3 \int_0^t\int_0^u\int_0^v\int_0^s f^2(t_1,s,v,u)dt_1 dsdvdu \right. \tag{9}$$

$$\left. -16v_0\rho^3 \int_0^t\int_0^u s\int_0^s f^2(t_1,s,u)dt_1 dsdu - 4v_0\rho^3 \int_0^t u\int_0^u\int_0^s f^2(t_1,s,u)dt_1 dsdu - 20v_0\rho^3 \int_0^t s^2\int_0^s f^2(t_1,s)dt_1 ds\right] f(t_1)$$

$$2\dot{v}_3^{(2)}(t;t_1,t_2) + 2v_0^{(1)}(t;t_1)v_2^{(1)}(t;t_2) + 2v_2^{(1)}(t;t_1)v_0^{(1)}(t;t_2) + 2v_1^{(1)}(t;t_1)v_1^{(1)}(t;t_2)$$

$$+ 8\int_0^t v_1^{(2)}(t;t_1,t_3)v_1^{(2)}(t;t_2,t_3)dt_3 + 4v_0^{(0)}(t)v_2^{(2)}(t;t_1,t_2) + 4v_1^{(0)}(t)v_1^{(2)}(t;t_1,t_2)$$

$$+12\int_0^t v_0^{(1)}(t;t_3)v_2^{(3)}(t;t_1,t_2,t_3)dt_3 = 0,$$

$$v_3^{(2)}(t;t_1,t_2) = \left[-6v_0^2\rho^2 t^3 - 10\rho^4\int_0^t s^2\int_0^s f^2(t_1,s)dt_1 ds - 2\rho^4\int_0^t u\int_0^u\int_0^s f^2(t_1,s,u)dt_1 dsdu\right.$$

$$\left.-4\rho^4\int_0^t\int_0^u\int_0^v\int_0^s f^2(t_1,s,v,u)dt_1 dsdvdu - 8\rho^4\int_0^t\int_0^u s\int_0^s f^2(t_1,s,u)dt_1 dsdu\right]f(t_1)f(t_2),$$

$$6v_3^{(3)}(t;t_1,t_2,t_3)+12v_0^{(0)}(t)v_2^{(3)}(t;t_1,t_2,t_3)+12v_0^{(1)}(t;t_1)v_2^{(2)}(t,t_2,t_3)+12v_1^{(1)}(t;t_1)v_1^{(2)}(t;t_2,t_3)=0,$$

$$v_3^{(3)}(t;t_1,t_2,t_3) = -4v_0\rho^3 t^3 f(t_1)f(t_2)f(t_3),$$

$$24v_3^{(4)}(t;t_1,t_2,t_3,t_4)+24v_1^{(2)}(t;t_1,t_2)v_1^{(2)}(t;t_3,t_4)+48v_0^{(1)}(t;t_4)v_2^{(3)}(t;t_1,t_2,t_3) = 0.$$

$$v_3^{(4)}(t,t_1,t_2,t_3,t_4) = -\rho^4 t^3 f(t_1)f(t_2)f(t_3)f(t_4).$$

PICARD APPROXIMATION

It is based on generating recursive convergent approximations for the solution until reaching certain level of accuracy defined according to the nature of the problem or to the measure in the scope of the analysis [19,20].

In the case of nonlinear Langevin's equation, Picard algorithm converges to the exact instant of the time at which the solution explodes. In this case there is a relation between the number of approximations and the values of parameters. The more number of approximations, the closer we are to the exact solution.

First, an arbitrary initial approximation should be suggested $v_{(0)}(t;\omega)$, and the numerical recursive equation will be [21]:

$$\upsilon_{(k)}(t_j;\omega) = \upsilon_{(0)}(t_j;\omega) - \varepsilon\sum_{i=0}^{j}\upsilon_{(k-1)}^2(t_i;\omega)\Delta t + \rho\sum_{i=0}^{j}f(t_i)\Delta W_{t_{i+1}}, \quad j=0,1,\cdots,N \tag{10}$$

which is substituted in Equation (2). First let us discuss the convergence of the model taking and $f(t) = e^{-\lambda t}$ then the equation will be linear model in terms of Brownian motion as:

$$\frac{d\upsilon}{dt} = \rho e^{-\lambda t} n(t), \quad \upsilon(0) = v_0, \tag{11}$$

$$\upsilon(t)(t) = v_0 + \rho\int_0^t e^{-\lambda s} dW_s, \tag{12}$$

$$E(\upsilon(t)) = v_0$$

$$\sigma^2 = \frac{\rho^2}{2\lambda}(1 - e^{-2\lambda t}). \tag{13}$$

We shall choice two arbitrary initial approximations in Picard approximations to simulate the model of Langevin's equation, the two cases of initial approximations are:

$$\varphi_{(0)}(t;\omega) = v_0 \quad \text{and} \quad \varphi_{(0)}(t;\omega) = v_0 + \rho\sum_{i=0}^{j}f(t_i)\Delta W_{t_{i+1}}$$

HOMOTOPY PERTURBATION METHOD (HPM)

The HPM is proved to be an effective, simple, and convenient to solve nonlinear PDE problems [22,23]. In HPM, a parameter, p is embedded in a homotopy function which satisfies

$$H(\upsilon, p) = [R(\upsilon) - R(u_0)] + p[A(\upsilon) - F(r)] = 0, \tag{14}$$

where u_0 is an initial approximation to the solution of the equation:

$$A(\upsilon) - F(r) = 0, \quad r \in \phi, \tag{15}$$

with the boundary conditions:

$$B\left(\upsilon, \frac{\partial \upsilon}{\partial n}\right) = 0, \ r \in \phi. \tag{16}$$

In which A is a nonlinear differential operator which can be decomposed into a linear operator R and a nonlinear operator N, B is a boundary operator and F(r) is a known analytical function. The homotopy introduces a continuously deformed solution for the case of $p = 0$, $R(\upsilon) - R(\upsilon_0) = 0$ is the original equation. The basic assumption of the HPM method is that the solution of the original equation can be expanded as a power series in p as,

$$\upsilon = \sum_{i=0}^{\infty} p^i \upsilon_i = \upsilon_0 + p\upsilon_1 + p^2 \upsilon_2 + p^3 \upsilon_3 + \cdots. \tag{17}$$

Now, setting p=1, the approximate solution is obtained as

$$\upsilon = \lim_{p \to 1} \upsilon = \sum_{i=0}^{\infty} \upsilon_i = \upsilon_0 + \upsilon_1 + \upsilon_2 + \upsilon_3 + \cdots. \tag{18}$$

The rate of convergence of HPM depends greatly on the initial approximation υ_0 which is considered as the main disadvantage of HPM [9-11]. The idea of the embedded parameter can be utilized to solve nonlinear problems by embedding this parameter to the problem and then forcing it to be unity in the obtained approximate solution if the convergence can be assured.

HPM is a simple technique which enables the extension of the applicability of the perturbation methods from small value applications to general ones. Applying HPM technique on Equation (2) with,

$$A(\upsilon) = R(\upsilon) + \varepsilon \upsilon^2, \tag{19}$$

$$R(\upsilon) = \frac{d\upsilon(t;\omega)}{dt}, \tag{20}$$

Solution of Nonlinear Stochastic Langevin's Equation Using

$$N(\upsilon) = \varepsilon \upsilon^2, \qquad (21)$$

$$F(r) = \rho f(t) n(t;\omega). \qquad (22)$$

The homotopy function takes the following form:

$$H(\upsilon,p) = (1-p)\left[R(\upsilon) - R(u_0)\right] + p\left[A(\upsilon) - f(r)\right] = 0, \qquad (23)$$

Equivalently:

$$R(\upsilon) - R(u_0) + p\left[R(v_0) + \varepsilon \upsilon^2 - \rho f(t) n(t)\right] = 0. \qquad (24)$$

Using Equation (17) in Equation (24) and equating the similar powers of p in both sides of the equation, one can get the following:

$$\upsilon_0 = u_0;\quad u_0(0) = v_0. \qquad (25)$$

$$R(\upsilon_1) = \rho f(t) n(t) - R(v_0) - \varepsilon v_0^2, \upsilon_1(0) = 0,$$

$$\upsilon_1(t) = \rho \int_0^t f(s) n(s) \, ds - \varepsilon v_0^2 t. \qquad (26)$$

$$R(\upsilon_2) = -2\varepsilon \upsilon_0 \upsilon_1,\quad \upsilon_2(0) = 0,$$

$$\upsilon_2(t) = -2\varepsilon \rho v_0 \int_0^t \int_0^s f(u) n(u) \, du\, ds + \varepsilon^2 v_0^3 t^2. \qquad (27)$$

$$R(\upsilon_3) = -\varepsilon \upsilon_1^2 - 2\varepsilon \upsilon_0 \upsilon_2,\quad \upsilon_3(0) = 0,$$

$$v_3(t) = -\varepsilon\rho^2 \int_0^t \left[\int_0^s f(u)n(u)du\right]^2 ds - \varepsilon^3 v_0^4 t^3 + 2\varepsilon^2 v_0^2 \rho \int_0^t s \int_0^s f(u)n(u)duds$$

(28)

$$+ 4\varepsilon^2 v_0^2 \rho \int_0^t \int_0^s \int_0^u f(\alpha)n(\alpha)d\alpha du ds.$$

$$R(v_4) = -2\varepsilon v_0 v_3 - 2\varepsilon v_1 v_2, \quad v_4(0) = 0,$$

$$v_4(t) = 2\varepsilon^2 v_0 \rho^2 \int_0^t \int_0^s \left[\int_0^u f(\alpha)n(\alpha)d\alpha\right]^2 du ds - 4\varepsilon^3 v_0^3 \rho \int_0^t \int_0^s u \int_0^u f(\alpha)n(\alpha)d\alpha du ds$$

$$-8\varepsilon^3 v_0^3 \rho \int_0^t \int_0^s \int_0^u \int_0^v f(\alpha)n(\alpha)d\alpha dv du ds - 2\varepsilon^3 v_0^3 \rho \int_0^t s^2 \int_0^s f(u)n(u)du ds + \varepsilon^4 v_0^5 t^4$$

(29)

$$-4\varepsilon^3 v_0^3 \rho \int_0^t s \int_0^s \int_0^u f(\alpha)n(\alpha)d\alpha du ds + 4\varepsilon^2 v_0 \rho^2 \int_0^t \int_0^s f(\alpha)n(\alpha)\left[\int_0^s \int_0^u f(\beta)n(\beta)d\beta du\right] d\alpha \right] ds.$$

We have many choices in guessing the initial approximation together with its initial conditions which greatly affects the consequent approximations. We shall consider the choice of

$$v_0(t) = v_0,$$

(30)

which satisfies the initial condition. The statistical measures of the different orders can be obtained as:

1st order:

$$v^{(1)}(t) = v_0 + v_1,$$

Solution of Nonlinear Stochastic Langevin's Equation Using

$$\mu_{v^{(1)}}(t;\omega) = v_0(1-\varepsilon v_0 t),$$

$$\sigma^2_{v^{(1)}} = \rho^2 \int_0^t f^2(s)\,ds, \qquad (31)$$

2nd order:

$$v^{(2)}(t) = v^{(1)}(t) + v_2,$$

$$\mu_{v^{(2)}}(t;\omega) = v_0\left(1-\varepsilon v_0 t + \varepsilon^2 v_0^2 t^2\right), \qquad (32)$$

$$\sigma^2_{v^{(2)}} = \rho^2 \int_0^t f^2(s)\,ds - 4\varepsilon\rho^2 v_0 t \int_0^t f^2(u)\,ds + 4\varepsilon^2 \rho^2 v_0^2 t \int_0^t \int_0^s f^2(u)\,du\,ds,$$

3rd order:

$$v^{(3)}(t) = v^{(2)}(t) + v_3,$$

$$\mu_{v^{(3)}}(t;\omega) = v_0\left(1-\varepsilon v_0 t + \varepsilon^2 v_0^2 t^2 - \varepsilon^3 v_0^3 t^3\right) - \varepsilon\rho^2 \int_0^t \int_0^s f(u)^2\,du\,ds,$$

$$\sigma^2_{v^{(3)}} = \rho^2 \int_0^t f^2(s)\,ds + 4\varepsilon^2 \rho^2 v_0^2 t \int_0^t \int_0^s f^2(u)\,du\,ds$$

$$-4\varepsilon\rho^2 v_0 t \int_0^t f^2(s)\,ds + 6\varepsilon^2 v_0^2 \rho^2 t^2 \int_0^t f^2(s)\,ds \qquad (33)$$

$$+6\varepsilon^4 v_0^4 \rho^2 t^2 \int_0^t s \int_0^s f^2(u)\,du\,ds + 2\varepsilon^2 \rho^4 t \left[\int_0^t \int_0^s f^2(u)\,du\,ds\right]^2$$

$$-12\varepsilon^3 v_0^3 \rho^2 t^2 \int_0^t\int_0^s f^2(\alpha)\,d\alpha ds + 12\varepsilon^4 v_0^4 \rho^2 t^2 \int_0^t\int_0^s\int_0^u f^2(\alpha)\,d\alpha du ds,$$

4th order:

$$\upsilon^{(4)}(t) = \upsilon^{(3)}(t) + \upsilon_4,$$

$$\mu_{\upsilon^{(4)}}(t;\omega) = v_0\left(1 - \varepsilon v_0 t + \varepsilon^2 v_0^2 t^2 - \varepsilon^3 v_0^3 t^3 + \varepsilon^4 v_0^4 t^4\right) - \varepsilon \rho^2 \int_0^t\int_0^s f(u)^2\,du ds$$

$$+ 2\varepsilon^2 v_0 \rho^2 \int_0^t\int_0^s\int_0^u f(\alpha)^2\,d\alpha du ds + 4\varepsilon^2 v_0 \rho^2 \int_0^t s \int_0^s f(u)^2\,du ds,$$

$$\sigma^2_{\upsilon^{(4)}} = \sigma^2_{\upsilon^{(3)}} + 8\varepsilon^6 v_0^6 \rho^2 t^3 \int_0^t s^2 \int_0^s f^2(\alpha)\,d\alpha ds + 32\varepsilon^6 v_0^6 \rho^2 t^3 \int_0^t\int_0^s\int_0^u\int_0^v f^2(\alpha)\,d\alpha dv du ds$$

$$+ 16\varepsilon^6 v_0^6 \rho^2 t^3 \int_0^t s \int_0^s \int_0^u f^2(\alpha)\,d\alpha du ds + 12\varepsilon^4 v_0^2 \rho^4 t^2 \int_0^t\int_0^s \left[\int_0^u f^2(\alpha)\,d\alpha\right]^2 du ds$$

(34)

$$+ 24\varepsilon^4 v_0^2 \rho^4 t^2 \int_0^t\int_0^s f^2(\alpha) \int_0^s\int_0^u f^2(\beta)\,d\beta du d\alpha ds + 16\varepsilon^6 v_0^6 \rho^2 t^3 \int_0^t\int_0^s u \int_0^u f^2(\alpha)\,d\alpha du ds$$

$$- 8\varepsilon^3 v_0^3 \rho^2 t^3 \int_0^t f^2(\alpha)\,d\alpha + 16\varepsilon^4 v_0^4 \rho^2 t^3 \int_0^t\int_0^s f^2(\alpha)\,d\alpha ds - 6\varepsilon^5 v_0^5 \rho^2 t^3 \int_0^t s \int_0^s f^2(\alpha)\,d\alpha ds$$

$$- 32\varepsilon^5 v_0^5 \rho^2 t^3 \int_0^t\int_0^s\int_0^u f^2(\alpha)\,d\alpha du ds - 12\varepsilon^3 v_0 \rho^4 t^2 \int_0^t \left[\int_0^s f^2(\alpha)\,d\alpha\right]^2 ds.$$

RESULTS

Results of WHEP Technique

Figures 1-8 show the mean and the variance of fourth order for first, second and third corrections. As shown in Figure 1 for case of $\varepsilon = 0.5$ there is fluctuation of the solution to positive and negative infinity (first and third corrections to negative and second to positive) we can guess that the fourth correction will diverge to positive infinity, this guess will be proved later in this paper but in Figure 2 for case of $\varepsilon = 0.01$, the three corrections converge in t<0.5, the second and the third corrections will deviate from first one.

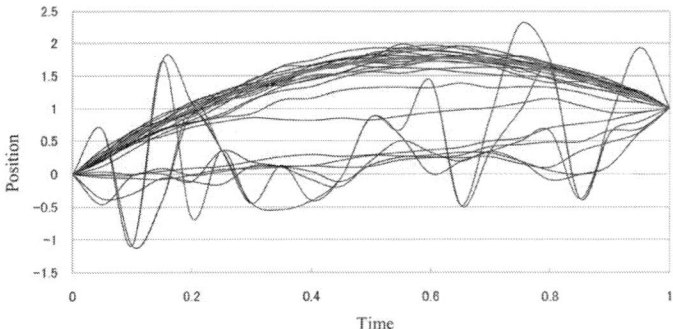

Figure 1: Fourth order mean of first, second and third corrections. Comparison between the different corrections for $\varepsilon = 0.5$ of pure white noise.

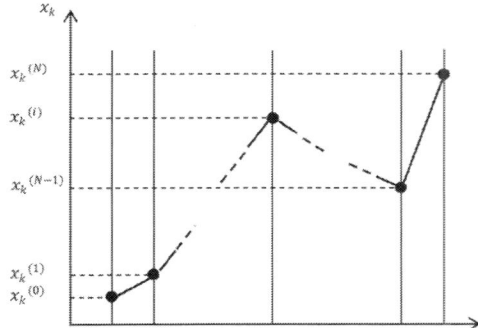

Figure 2: Fourth order variance of first, second and third corrections. Comparison between the different corrections for $\varepsilon = 0.5$ of pure white noise.

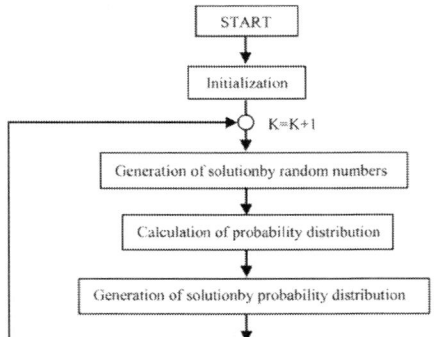

Figure 3: Fourth order mean of first, second and third corrections. Comparison between the different corrections for $\varepsilon = 0.01$ of pure white noise.

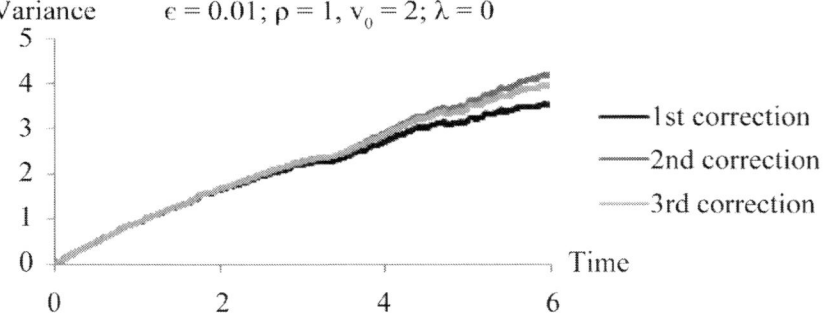

Figure 4: Fourth order variance of first, second and third corrections. Comparison between the different corrections for $\varepsilon = 0.01$ of pure white noise.

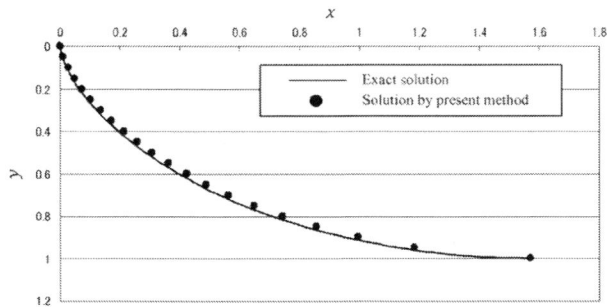

Figure 5: Fourth order mean of first, second and third corrections. Comparison between the different corrections for $\varepsilon = 0.5$ of decaying white noise.

Solution of Nonlinear Stochastic Langevin's Equation Using

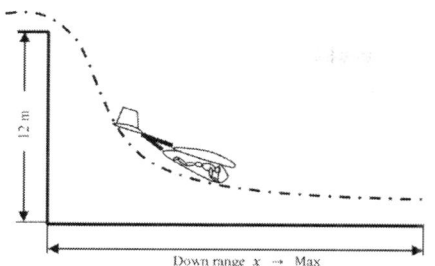

Figure 6: Fourth order variance of first, second and third corrections. Comparison between the different corrections for $\varepsilon = 0.5$ of decaying white noise.

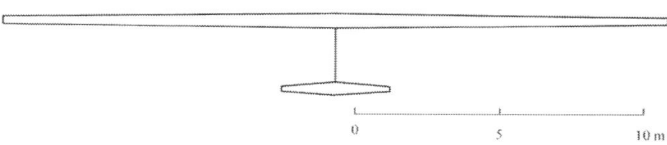

Figure 7: Fourth order mean of first, second and third corrections. Comparison between the different corrections for $\varepsilon = 0.01$ of decaying white noise.

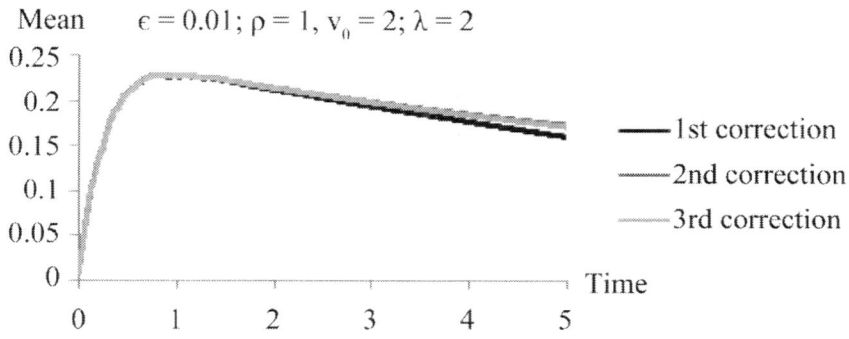

Figure 8: Fourth order variance of first, second and third corrections. Comparison between the different corrections for $\varepsilon = 0.01$ of decaying white noise.

For small value of nonlinearity strength $\varepsilon=0.01$, the divergence of solution which proved in [8] using the growth condition occurred in later interval such as in Figures 3, 4, 7 and 8 was occurred after t = 3 and in Figures 1, 2, 5 and 6 for $\varepsilon=0.5$, the divergence was occurred nearly at t=0.5.

There is no obvious difference in mean and variance for case of $\lambda=0$ or $\lambda=2$ in Figures 1, 2, 5 and 6. In Figures 2 and 6 represent the convergence of all corrections for $\varepsilon=0.5$ until instant of time then the divergence of the solution begins; this convergence proves that for some higher corrections or orders we can obtain accurate solution and know at which instant of time the divergence exactly happens.

Figures 6 and 8 for $\lambda=2$, decreasing in the magnitude of the variance has been occurred, we can conclude that increasing the value of λ, somehow works on eliminating the effect of the unknown pole which causes the explosion of solution.

Decreasing the nonlinearity strength in Figures 6 and 8 make the variance increasing with time until fixed to certain time after which it begin to make the explosion.

To know the effect of increasing order in the accurate of the solution, fixed the third correction as case of test as follows.

Figure 9 shows the mean in different models which give us knowledge about decreasing ε which make the divergence very slowly and delay it to long instant of time but the question is what is the value of this parameter until the system doesn't make explosion? from the study, decreasing the nonlinearity strength need long interval of time to know the behavior of the solution, for $\varepsilon=0.5$ and $\lambda=0;2$ the mean deceasing with time until reach to zero and continued its trajectory to negative infinity and the effect of λ is the same for delaying the explosion, but the question is, which of this parameters is stronger?. From previous models the nonlinearity strength has the strong effect in mean and variance but this doesn't decay the strong effect of λ in variance as said before.

Solution of Nonlinear Stochastic Langevin's Equation Using

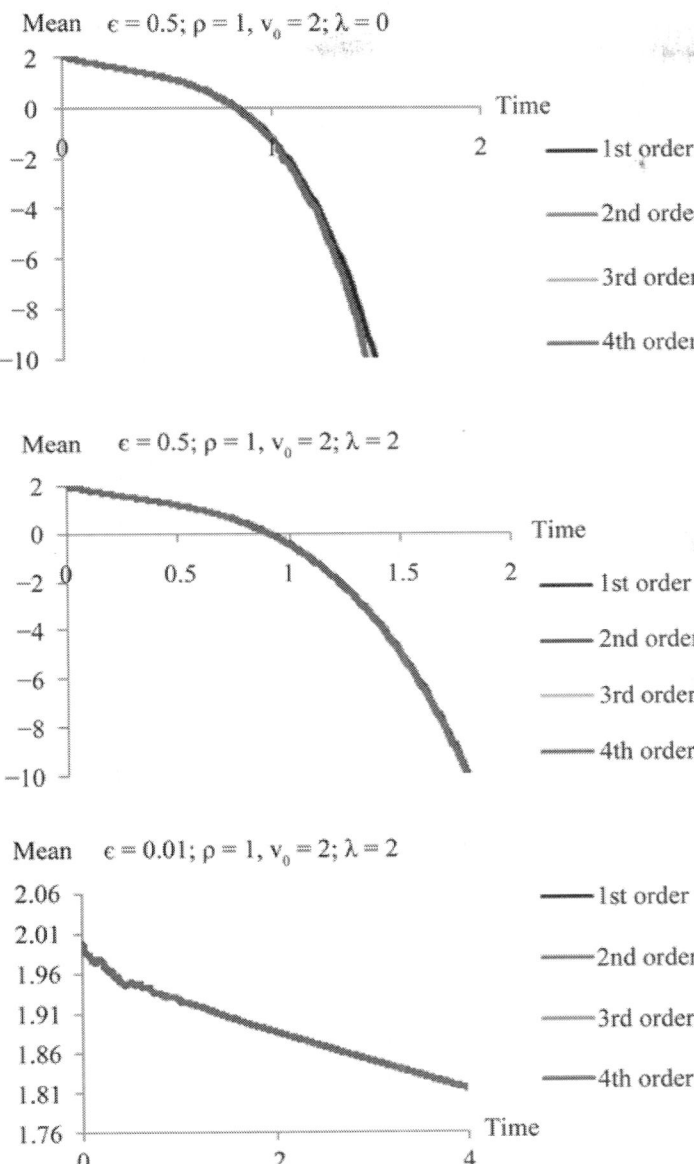

Figure 9: Third correction mean of first, second, third and fourth orders. Comparison between the different orders for $\varepsilon = 0.5$, $\varepsilon = 0.01$. Case of pure and decaying white noise.

Figures 10 shows the coincident of all orders such as that in mean but in this case of the variance, we find the first order deviate from the others at increasing time and the interval of that depend on the parameter's value, all plots of variance has the same trajectory of increasing variance to reach its peak then decreasing to certain time after which it make the deviation.

Results of Picard Approximation

Figures 11-14 prove that Picard approximation gives good result for $\varepsilon < 1$. The two cases of arbitrary initial approximations have almost the same results in mean and in variance until instant of time where the deviation occurred as follows in Figures 15-18 for $\varepsilon = 0.5; 0.01$ which mean we need to decrease ε and there was a convergence between the two cases of initial approximation in the results of Picard's approximation.

Results of HPM Method

Figures 5 and 6 are semi-similar to Figures 19 and 20 at which we can see the simulations of the orders in HPM are look like the simulations of the corrections in WHEP technique and the fourth order here is similar to our guess of fourth correction which diverge to positive infinity.

Figures 7 and 8 are semi-similar to Figures 21 and 22 which prove that increasing order of solution will give the accurate location at which the solution explodes and more convergence to the exact solution of this model.

Solution of Nonlinear Stochastic Langevin's Equation Using

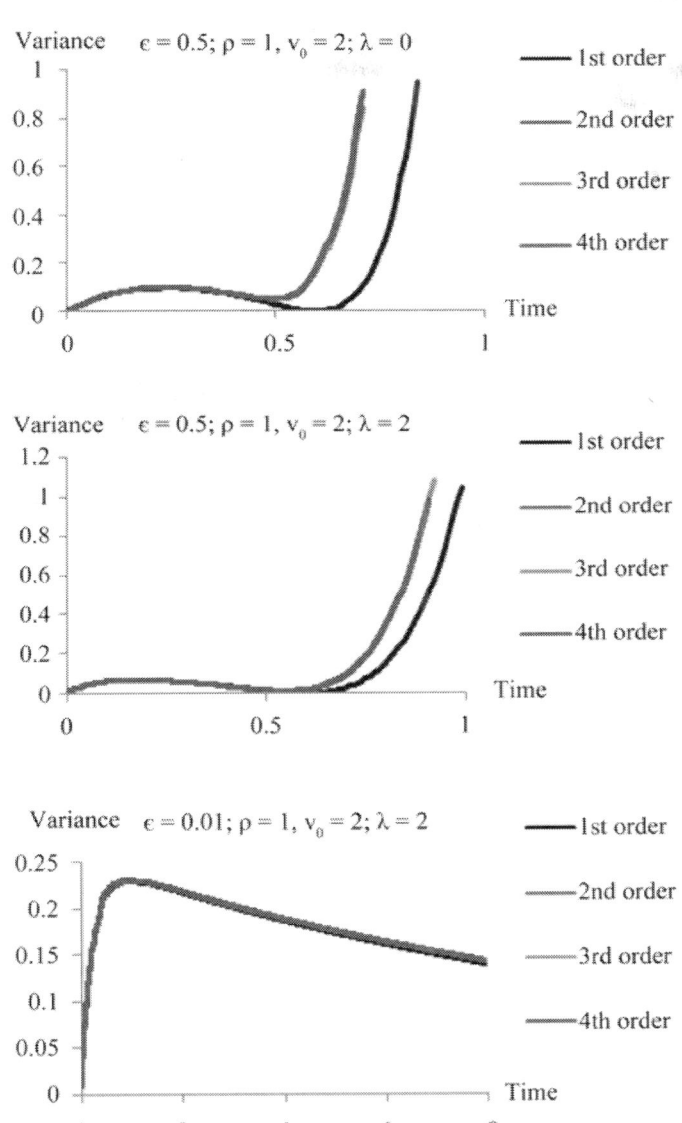

Figure 10: Third correction variance of first, second, third and fourth orders. Comparison between the different orders for $\varepsilon = 0.5$, $\varepsilon = 0.01$. Case of pure and decaying white noise.

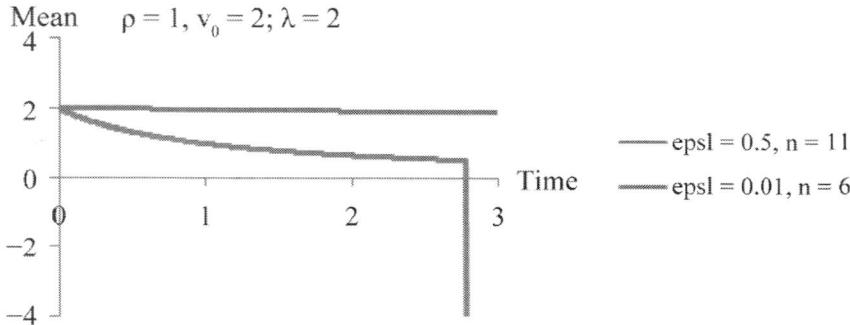

Figure 11: Mean of last iteration (n) for $\varepsilon = 0.5$; 0.01 of first case of initial approximation.

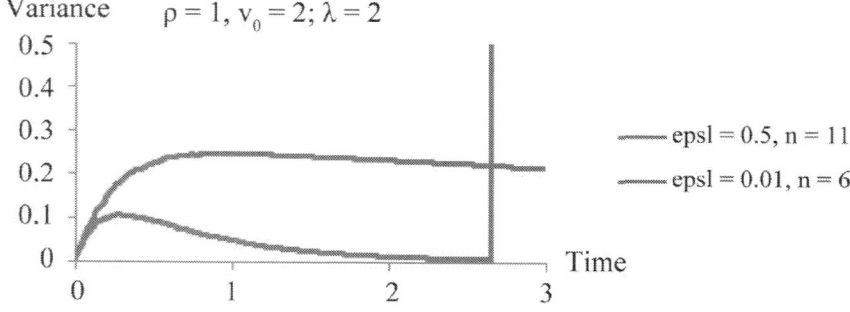

Figure 12: Variance of last iteration (n) for $\varepsilon = 0.5$; 0.01 of first case of initial approximation.

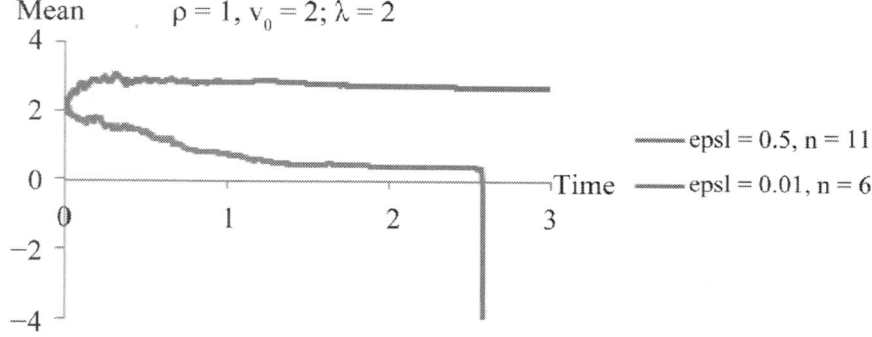

Figure 13: Mean of last iteration (n) for $\varepsilon = 0.5$; 0.01 of second case of initial approximation.

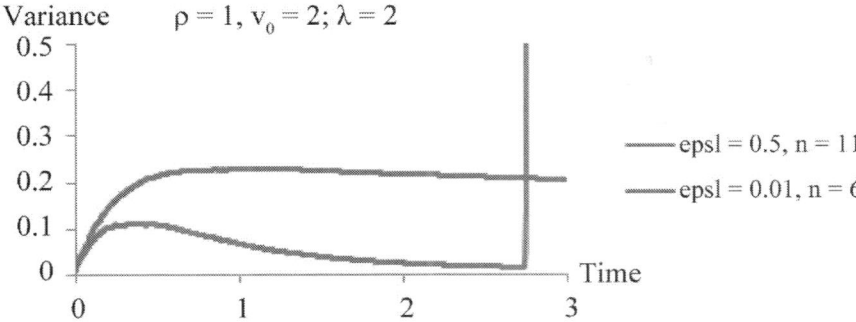

Figure 14: Variance of last iteration (n) for ε = 0.5; 0.01 of second case of initial approximation.

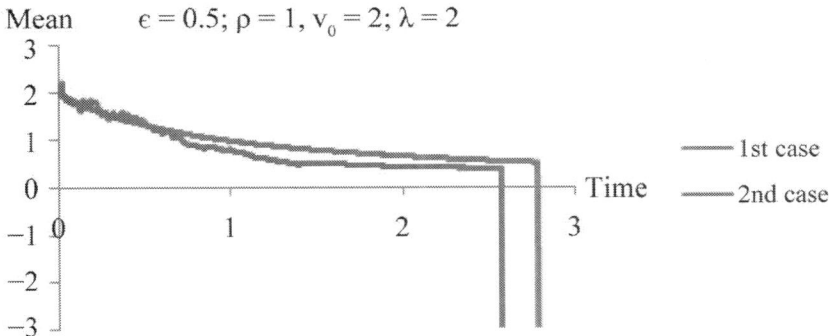

Figure 15: Comparison between Picard mean using the first and second cases of initial approximations for ε = 0.5.

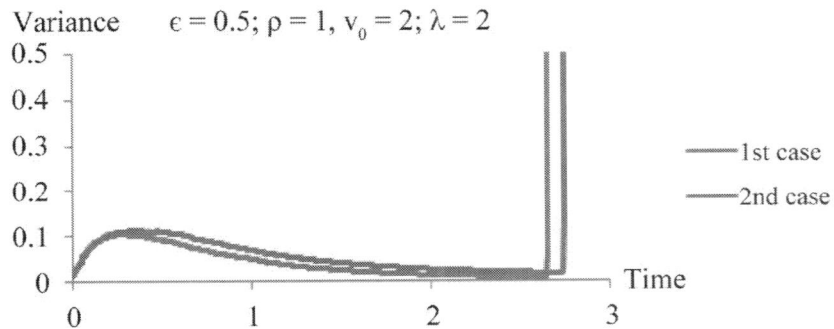

Figure 16: Comparison between Picard variance using the first and second cases of initial approximations for ε = 0.5.

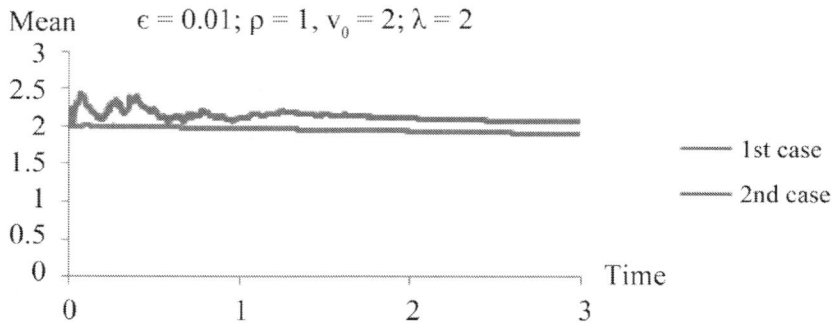

Figure 17: Comparison between Picard mean using the first and second cases of initial approximations for $\varepsilon = 0.01$.

The changing of ε makes a difference in the results, in Figure 23 increasing ε than 1 doesn't make any prob- lem in the solution but decreasing this value gives good results, and also changing in ρ doesn't make any change in the mean, but in variance it makes change in delaying the explosion and gives good solution especially for decreasing this parameter.

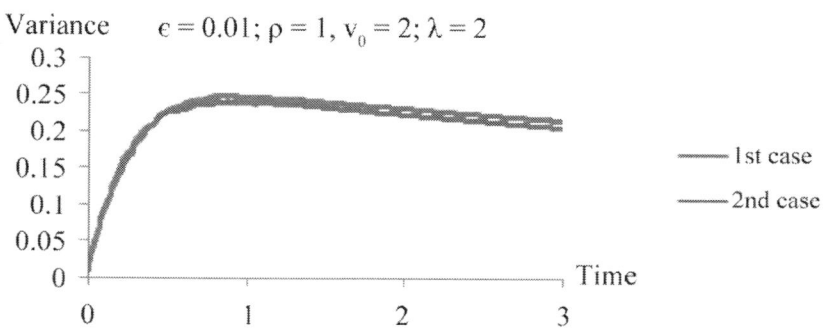

Figure 18: Comparison between Picard variance using the first and second cases of initial approximations for $\varepsilon = 0.01$.

Solution of Nonlinear Stochastic Langevin's Equation Using

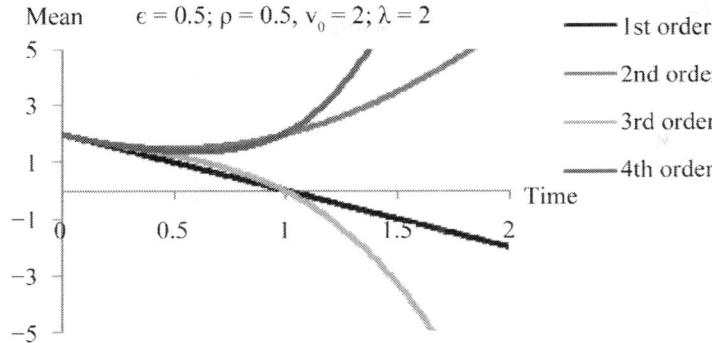

Figure 19: Comparison of the first, second, third and fourth orders mean for $\varepsilon = 0.5$ and $\lambda = 2$.

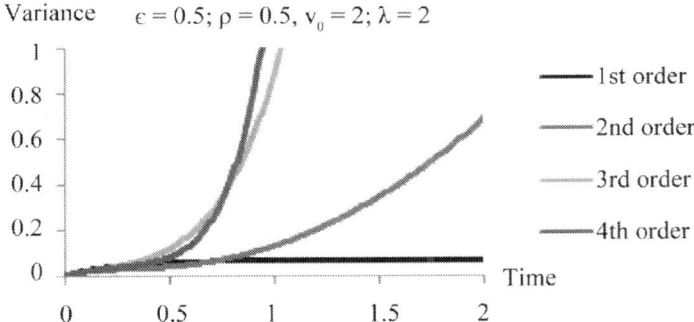

Figure 20: Comparison of the first, second, third and fourth orders variance for $\varepsilon = 0.5$ and $\lambda = 2$.

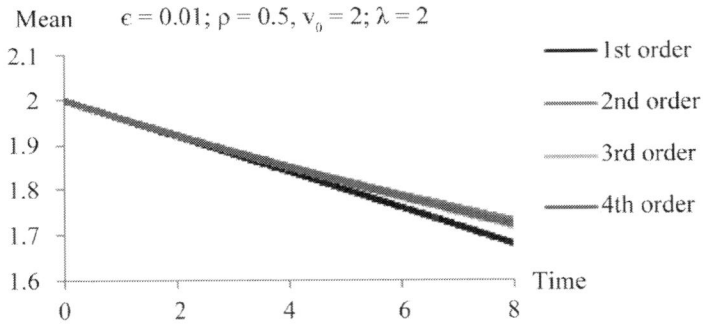

Figure 21: Comparison of the first, second, third and fourth orders mean for $\varepsilon = 0.01$ and $\lambda = 2$.

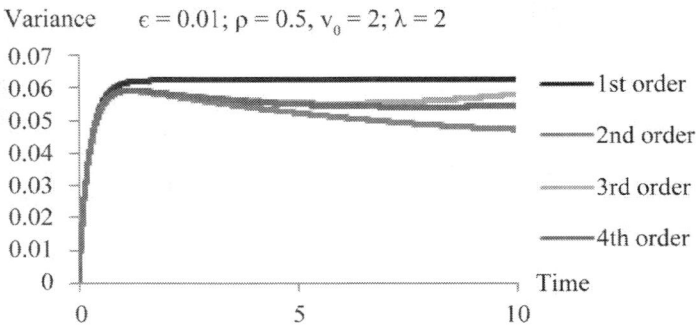

Figure 22: Comparison of the first, second, third and fourth orders variance for $\varepsilon = 0.01$ and $\lambda = 2$.

Comparison between Different Methods

The comparison between all methods depends greatly on decreasing the nonlinearity strength (ε), decreasing this value gives more convergence, Figures 24 and 25 show the convergence between the three methods for first case of Picard initial approximation, although the convergence between the three methods for the second case of Picard initial approximation is less than the first one, because of the increasing of the term of white noise Picard in the second case of Picard. We can deduce that the first case of initial approximation make Picard approxi- mation technique converges to WHEP and HPM techniques. It is worth in Figures 24 and 25 to note that, Picard n = 6 means using Picard iterate up to the 6th approximation. Also WHEP 4, 3 means using the WHEP technique with 4th order and 3rd correction. HPM 4 means using the 4th order of HPM method.

CONCLUSIONS

Picard is reaching the solution in a numerical way to deduce the mathematical formulas which deplete too much effort despite of the convergence between the two cases of initial approximations: the first case of initial appro- ximation was more convergent to WHEP technique and

HPM methods than the second one. The WHEP technique seems to be efficient because of its corrections in spite of being analytically lengthy. The HPM is easier in computation but for higher order to calculate the variance, and it is complicated but not more than the WHEP which needs huge analytical calculations and formulations. HPM expectedly depends highly on the initial guess. WHEP and HPM are convergent to each other than any other two compared methods, and every order in HPM has the same trajectory of the opposite correction in WHEP, which means the first order in HPM looks like the first correction in WHEP and so until the fourth one.

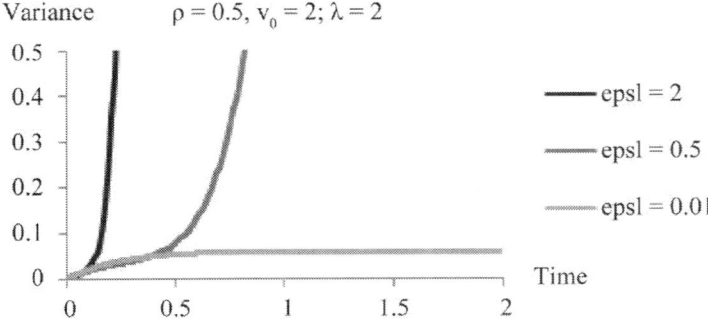

Figure 23: The fourth order variance for different ε and $\lambda = 2$.

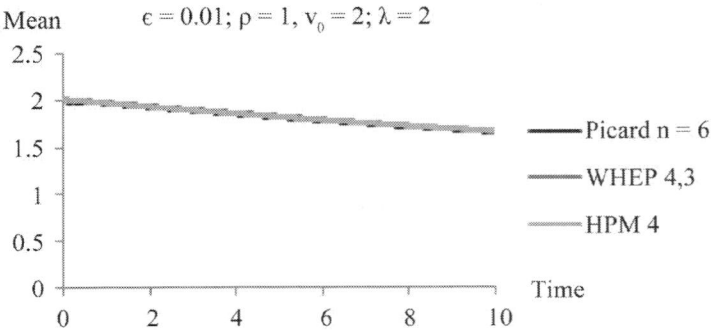

Figure 24: Comparison between Picard, WHEP and HPM fourth order mean for first case of Picard.

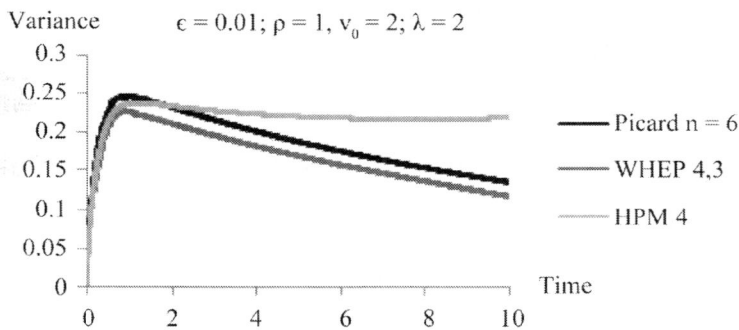

Figure 25: Comparison between Picard, WHEP and HPM fourth order variance for first case of Picard.

REFERENCES

1. K. De Feriet, "Random Solutions of Partial Differential Equations," Proceedings of the 3rd Berkeley Symposium on Mathematical Statistics and Probability, Vol. III, 1956, pp. 199-208.
2. R. Bharucha, "A Survey on the Theory of Random Functions," The Institute of Mathematical Sciences, Matscience Report 31, 1965.
3. V. Lo Dato, "Stochastic Processes in Heat and Mass Transport," Probabilistic Methods in Applied Mathematics, Vol. 3(A), 1973, pp. 183-212.
4. B. A. Georges, "Random Generalized Solutions to the Heat Equations," Journal of Mathematical Analysis and Applications, Vol. 60, No. 1, 1977, pp. 93-102. http://dx.doi.org/10.1016/0022-247X(77)90051-8
5. W. T. Coffey and Y. P Kalmykov, "The Langevin Equation, with Applications to Stochastic Problems in Physics, Chemistry and Electrical Engineering," 3rd Edition, World Scientific Publishing Company, Singapore City, 2012.
6. M. A. El-Tawil, "The Application of the WHEP Technique on Partial Differential Equations," International Journal of Differential Equations and Applications, Vol. 7, No. 3, 2003, pp. 325-337.
7. M. A. El-Beltagy and M. A. El-Tawil, "Toward a Solution of a Class of Non-Linear Stochastic perturbed PDEs Using Automated WHEP Algorithm," Applied Mathematical Modelling, 2013, in Press. http://dx.doi.org/10.1016/j.apm.2013.01.038
8. M. El Tawil and W. Shawky "Wiener Functional, Integrals, and Stochastic Differential Equations Solutions Using Euler- Maruyama, Picard and WHEP Computer Simulation Stud," BSc Thesis, Engineering faculty, Cairo University, Cairo, 2011.
9. M. A. El-Tawil and A. Fareed, "Solution of Stochastic Cubic and Quintic Non-linear Diffusion Equation Using WHEP, Pickard and HPM Methods," Open

Journal of Discrete Mathematics, Vol. 1, No. 1, 2011, pp. 6-21. http://dx.doi.org/10.4236/ojdm.2011.11002
10. M. El-Tawil and N. El-Molla, "The Approximate Solution of a Nonlinear Diffusion Equation Using Some Techniques, a Comparison Study," Applied Mathematics and Computing, Vol. 29, No. 1-2, 2009, pp. 281-299.
11. A. S. El-Johani, "Comparisons between WHEP and Homotopy Perturbation Techniques in Solving Stochastic Cubic Oscillatory Problems," AIP Conference Proceedings, Vol. 1148, 2010, pp. 743-752. http://dx.doi.org/10.1063/1.3225426
12. J. C. Cortes, J. V. Romero, M. D. Rosello and R. J. Villanueva, "Applying the Wiener-Hermite Random Technique to Study the Evolution of Excess Weight Population in the Region of Valencia (Spain)," American Journal of Computational Mathe- matics, Vol. 2, No. 4, 2012, pp. 274-281. http://dx.doi.org/10.4236/ajcm.2012.24037
13. N. Wiener, "Nonlinear Problems in Random Theory," MIT Press, John Wiley, Cambridge, 1958.
14. R. H. Cameron and W. T. Martin, "The Orthogonal Development of Non-Linear Functionals in Series of Fourier-Hermite Functionals," Annals of Mathematics, Vol. 48, No. 2, 1947, pp. 385-392. http://dx.doi.org/10.2307/1969178
15. T. Imamura, W. Meecham and A. Siegel, "Symbolic Calculus of the Wiener Process and Wiener-Hermite Functionals," Jour- nal of Mathematical Physics, Vol. 6, No. 5, 1965, pp. 695-706. http://dx.doi.org/10.1063/1.1704327
16. W. C. Meecham and D. T. Jeng, "Use of the Wiener-Hermite Expansion for Nearly Normal Turbulence," Journal of Fluid Mechanics, Vol. 32, No. 2, 1968, pp. 225-235. http://dx.doi.org/10.1017/S0022112068000698
17. R. Riganti and N. Bellomo, "Nonlinear Stochastic Systems in Physics and Mechanics," World Scientific Publishing Co., Singapore City, 1987.
18. M. El-Beltagy and A. Al-Johani, "Higher-Order WHEP Solutions of Quadratic Nonlinear Stochastic Oscillatory Equation," Engineering, Vol. 5, No. 5A, 2013, pp. 57-69.
19. E. A. Ibijola and B. J. Adegboyegun, "A Comparison of Adomian's Decomposition Method and Picard Iterations Method in Solving Nonlinear Differential Equations," Nigeria Global Journal of Science Frontier Research Mathematics and Decision Sciences, Vol. 12, No. 7, 2012.
20. I. K. Youssef, "Picard Iteration Algorithm Combined with Gauss-Seidel Technique for Initial Value Problems," Applied Mathematics and Computation, Vol. 190, No. 1, 2007, pp. 345-355. http://dx.doi.org/10.1016/j.amc.2007.01.058
21. R. Riganti and N. Bellomo, "Nonlinear Stochastic Systems in Physics and Mechanics," World Scientific Publishing Co., Singapore City, 1987.
22. J. H. He, "Application of Homotopy Perturbation Method to Nonlinear Wave Equations," Chaos Solitons Fractals, Vol. 26, No. 3, 2005, pp. 295-300. http://dx.doi.org/10.1016/j.chaos.2005.03.006
23. J. H. He, "Homotopy Perturbation Method for Solving Boundary Value Problems," Physics Letters A, Vol. 350, No. 1-2, 2006, pp. 87-88.

CITATION

M. Hamed, M. El-Twail, B. El-desouky and M. El-Beltagy, "Solution of Nonlinear Stochastic Langevin's Equation Using WHEP, Pickard and HPM Methods," *Applied Mathematics*, Vol. 5 No. 3, 2014, pp. 398-412. doi:10.4236/am.2014.53041.

Index

A

Accurate solution 65, 66, 67, 69, 70
Adomian decomposition method (ADM) 40
agent-based computational models 150, 151
Analytic method 58

B

Brownian motion process 23, 24, 25, 32

C

Chemical Langevin Equation (CLE) 3

F

functional analysis 23, 37

H

heterogeneous 150, 153, 164, 165
Homotopy perturbation method (HPM) 191
Hyperbolic 169, 170, 173, 185
Hyperbolic regime 185

L

Lebesgue integral 24, 30, 32

M

meansquare(MS) 99
Monte-Carlo simulation 167, 168

N

non-linear system 102
Number of nonzero (NNZ) 173, 174
numerical methods 87, 88, 94
numerical simulation 101, 102

O

Optimal problem 68
Optimal solution 65, 66, 67, 68, 69, 70, 73, 77, 81, 84

P

Perturbation technique 54
Polynomial chaos expansion (PCE) 168

R

random integral equations 23
Random system 169

S

Simulated annealing (SA) 66, 68
Spectral stochastic finite element method (SSFEM) 168
Stirling number 149
Stochastic differential equations (SDE) 3
stochastic integral 23, 24, 25, 29, 30, 32, 34, 37
Stochastic linear system 177
stochastic mathematical models 163
Stochastic process optimization technique (SPOT) 65, 67
stochastic Runge-Kutta method (SRK) 87
Stratonovich-Taylor series 90
sufficiently derivative 89, 94

T

Temperature parameter 68

W

Wiener-Chaos expansion (WCE) 192
Wiener-Hermite expansion with perturbation (WHEP) 39, 191